SPEEDPRO SERIES

HOW TO POWER TUNE
MIDGET
SPRITE
FOR ROAD & TRACK

Daniel Stapleton

VELOCE PUBLISHING PLC
PUBLISHERS OF FINE AUTOMOTIVE BOOKS

SPEEDPRO SERIES

Other Veloce publications -

SpeedPro Series
How to Give Your MGB V-8 Power
by Roger Williams
*How to Build & Power Tune Weber DCOE
& Dellorto DHLA Carburetors*
by Des Hammill
*How to Power Tune the MGB 4-Cylinder
Engine*
by Peter Burgess
*How to Power Tune Alfa Romeo Twin Cam
Engines*
by Jim Kartalamakis

General
Alfa Romeo Owner's Bible
by Pat Braden
Alfa Romeo Modello 8C 2300
by Angela Cherrett
Alfa Romeo Giulia Coupe GT & GTA
by John Tipler
*Bubblecars & Microcars Colour Family
Album*
by Andrea & David Sparrow
Bugatti 46/50 - The Big Bugattis
by Barrie Price
Bugatti 57 - The Last French Bugatti
by Barrie Price

Car Bodywork & Interior: Care & Repair
by David Pollard
Chrysler 300
by Robert Ackerson
Citroen 2CV Colour Family Album
by Andrea & David Sparrow
Cobra - The Real Thing!
by Trevor Legate
*Completely Morgan: Three-Wheelers
1910-1952*
by Ken Hill
*Completely Morgan: Four-Wheelers 1936-
1968*
by Ken Hill
*Completely Morgan: Four-Wheelers from
1968*
by Ken Hill
Daimler SP250 'Dart'
by Brian Long
Fiat & Abarth 124 Spider & Coupe
by John Tipler
Fiat & Abarth 500 & 600
by Malcolm Bobbitt
Lola T70
by John Starkey
Making MGs
by John Price Williams
*Mazda MX5/Miata Enthusiast's Workshop
Manual*

by Rod Grainger & Pete Shoemark
Mini Cooper - The Real Thing!
by John Tipler
Morris Minor, The Secret Life of the
by Karen Pender
Motorcycling in the '50s
by Jeff Clew
Nuvolari: When Nuvolari Raced ...
by Valerio Moretti
Pass the MoT
by David Pollard
Porsche 911 R, RS & RSR
by John Starkey
*Rover P4 Series
(60•75•80•90•95•100•105•110)*
by Malcolm Bobbitt
Singer of Coventry
by Kevin Atkinson
*Triumph Motorcycles & The Meriden
Factory*
by Hughie Hancox
Triumph TR6
by William Kimberley
Vespa Colour Family Album
by Andrea & David Sparrow
Volkswagen Karmann Ghia
by Malcolm Bobbitt

First published in 1996 by Veloce Publishing Plc., Godmanstone, Dorset DT2 7AE, England. Fax 01300 341065.

ISBN 1 874105 68 5

Readers with ideas for automotive books, or books on other transport or related hobby subjects, are invited to write to Veloce Publishing at the above address.

British Library Cataloguing in Publication Data -
A catalogue record for this book is available from the British Library.

Typesetting (Soutane), design and page make-up all by Veloce on AppleMac.

Printed and bound in England.

Contents

Introduction & Acknowledgements

INTRODUCTION

The background to the writing of this book is my own MkIV Sprite which, when I bought it, had its wings (fenders) held on by rope! The car has swallowed more time and money than anyone - other than a fellow Midget/Sprite enthusiast - could imagine, yet it has been very reliable. I'm proud of my car and its capabilities even though close friends, after riding in it, have often been uncomplimentary and sometimes rude about it - and my driving!

In writing this book, my main aim has been to include as many different tuning/uprating projects as possible: all will be of interest and practical use to Sprite and Midget enthusiasts.

Where the project is a modification which has been carried out on my own car, it is explained in depth to illustrate particular points and likely pitfalls. Where the procedure for a particular project is covered in Midget/Sprite workshop manuals, such as replacement of standard front springs with uprated items, I have seen no need to write my own account of basic procedures, however, where the

modification is a job like fitting a bucket seat, I describe the fitting in some detail as I know of no other good source of practical guidance.

Anyone can pick up a random collection of performance parts and fit them to their car, but this is not likely to result in a satisfactory end product. I have found that some seemingly good tuning ideas do not always work as well as expected . . . Some modifications are best done in conjunction with others, or need some simple, but vital, finishing touch such as relief with a file. An example of this would be fitting Aldon anti-tramp bars, which need much minor modification if you already use, or plan to use, very wide tyres. The book covers all these hard-learned lessons to save you from the pitfalls I *almost* fell into - and some that I did!

Some of the modifications described do not consist of ready-made kits or parts that you can buy but are, instead, my own simple and original ideas. One such modification is the two-speed wiper mod, from which sprang the first ever article I wrote; this led to other articles and,

ultimately, this book. The two-speed wiper modification is not only one of the easiest to carry out, it's also extremely cheap and very, very useful.

My intention is that the information and guidance in this book will have a wide range of usefulness, giving as much practical help to those who just want to make minor improvements as to those seeking a start in motorsport. The book, I believe, has sufficient depth and usefulness to appeal to the home mechanic and a racing team.

This book will show you what tuning options there are, but choice of final specification will have to be yours. For instance, what I consider to be the ideal road or novice racer suspension set-up, another driver might find appalling or simply too ambitious for the road. Your choice, I hope, will be influenced by my own experiences and the equipment options and compatibility I have been able to discover. It's likely, too, that your finances will only permit limited modifications and, where appropriate, I indicate those which are necessities and those which are merely desirable.

A shot of the engine in the author's car during early ownership: most of the original tuning parts coming from his written-off 1300cc Mini. Gradually, he got the engine to produce reliable and decent horsepower.

Same car, same engine, but a few years later.

The book does not contain an extensive details of internal engine performance modifications. The reason for this is that David Vizard's *Tuning BL's A-Series Engine*, the definitive work on tuning the Sprite/Midget A-series engine already contains all the information you could need. It is, simply, the most authoritative book on the subject and, like a workshop manual, is more a necessity than an option if you're serious about getting to grips with your car. However, where I feel amplification on an engine-related matter is worthwhile, I provide this: an example is the conversion to DCOE Weber carburation which, at face value, seems relatively simple, but final choices on manifold and air filter can have far-reaching consequences, not least of which is trying to refit the bonnet!

Never give up on your own project and, should you ever have recourse to criticise this book, have information for inclusion in a future edition, or wish to seek my advice, I will be happy to hear from you - via a letter to the publisher.

I hope you'll find the book useful and recommend it to your friends.

ACKNOWLEDGEMENTS

I would like to acknowledge the help, advice, encouragement, assistance and prayers of various people who helped me put this book together, and, without whom, it would not have been possible.

Alan Abercrombie of Michelin Tyres plc, Alan (next door), Alan Allard, David Anton of APT, Brian Archer of Archer's Garage, Neil A Armstrong of Armteca Automotive Electronics, Julia Ashton of Corbeau Equipe Ltd, Simon Atherton, the Austin Healey Club, Peter Baldwin of Marshall's of Cambridge, Norman Barker of AP Racing, Geoff Barnes of AP Lockheed, Paul Beaurain of Pro-Align, Michael Bennett of SPA design, BGC, Roger Bishop of Serck Marston, Phil Bollen, Dan Boon of NGK Sparkplugs, Michael C Bown, (photographer), Matt Brazier, Karin Brinkmeier Hella KG, Alan Brown of Questmead Ltd (Mintex), John Bull, Pete Bulman, Keith Calver of MiniSpare, Ray Chadlee of AP, Ed Chan of The Project Company, John Chappel of Advanced Products (K & N Filters), Andy Chivers of RT Quaife Engineering Ltd, David Clarkson, Tom Colby of Speedwell Engineering, Peter Collen of AP Racing, Clare Collet of Metro Products Ltd, John Cracknell of Serck Marston, Alistair G. Curry of Mobil Oil Company Ltd, John Davenport of Lucas Aftermarket Operations, Rae Davis of Motobuild, George Demitriou of Minor Mania, Keith Dodd of Mini Spares, Jonathan Douglas of Induction Technology Group Ltd, Eight Clubs Ltd, Chris Eyles of Weber Concessionnaires, Tim Fenna of Frontline Spridget Ltd., John Fenning of Stockbridge Racing (Willans), Nessie Folwell of Corbeau Equipe Ltd, Rob Garofalo of Barrett Enterprises, Jim Garvey, Tony Gilhome of Dunlop, Mel Gosney of Serck Marston competition division, Steve Green of Goodyear Tyres, Eric Grundy, Geoff Hale (Austin Healey Club racer), Nick Hall of Safety Devices, Aaron Hammon of Victoria British Ltd, Nik Handford of Omni-Autos, Mr E Hankins of Mini Classic Cars, J Hansford of Luke, Brian Hill of BGH Geartech, Dave Hardy of Hardy Engineering, Derek Hibbert of Sport Photography, Mike Hodder of Spax, Pete Holden (Midget racer), Joe Huffaker of Huffaker Racing, Peter Huxley of Fuel System Enterprise (FSE), Jim Irwin of Pro-Align, Steve Jones (photographer), Nigel Kilerby of AVO/ Chassis Dynamics, Gail Latham of Serck Marston, Mark Lee of Lifeline, Simon A Lee of Janspeed, Mike Lennon of the Morris Minor Centre (Birmingham), Charles Leonard of Carcraft, Ashley Letts of Crane Cams, Madeleine and Russell Lightning of Lightning Studios, Alan Ling of T F Bell & Co (Insurance Brokers) Ltd, Richard Longman, John P MacPhail of Reco Prop, Ian Marr, James Martin Photos, Peter & Chris May of Peter May Engineering Ltd, Peter McBride of Ron Hopkinson MG Parts Centre, Bob McIvor of Microdynamics, the MG Owners Club, John Mulcock of Farndon Engineering, Mum, Sharon Munro (line drawings), Dave Musson of Speedograph Richfield Ltd, Mark Neil of Mintex Don, Davina Newsham of Unipart, Oselli Engineering Ltd, George Own of Goodridge, Simon Page (Austin Healey racer), Tim Parsley of Marshall's of Cambridge, Joyce Pearce, editor of Revcounter, John Peters of Sports & Classics, Tony Phillips of Kenlowe, Gordon Pluck of Marshall's of Cambridge, Robert Pocock of Kool Louvres, Andy Pringle of Castrol, Michael Quaife of RT Quaife Engineering Ltd, Donald Racine and others at Mini Mania, Steve Rae of Goodridge, Steve Renouf of Safety Devices, Dave Robinson (line drawings), Alan Rock of Stack Ltd, Paul 'Rodders' Rodman (1500 MG Midget racer), Phillip A Rogers (photographer), Steve Rollin of Northwest Import Parts, Richard E Rooks, Simon 'ex-RAF' Sinclair the Sprite & Midget Club, Jim Singmaster of Faspec, Pat Smith of Safety Devices, Phil Stapleton (my brother), Linda Stern of the BRSCC, Daniel Sturm of MK Parts, Kamaljit Tanda of the Morris Minor Centre

(Birmingham), Eric Taylor of Carr Reinforcements Limited, Charlotte Thoday, Mr S Tonks, Engineering Manager of Unipart International, the UK Techical Manager of Hella, David Vizard, Nick Warliker MIHORT, Simon 'ex-Army' Watson, Weber Concessionnaires Ltd, Mr P R Weston of Weston Body Hardware Limited, Helena Wilkinson, David Williams of Stack Ltd.

Daniel Stapleton
Northampton, England

A WORD FROM DAVID VIZARD

All too often practical books on performance tuning are written by those with little or no personal experience of actually doing the jobs described: when the reader tries to do the work, the writer's shortcomings soon become obvious!

Fortunately, Daniel Stapleton can put himself into the shoes of his readers so well that it seems as though he is turning the screwdrivers and pulling on the spanners and doing the very jobs he describes. His hands-on experience has taught him about the many small and unexpected problems and, of course, the real pitfalls that the reader - without such information - could easily fall foul of.

For all Midget/Sprite owners this book represents good solid 'how to' step-by-step instructions, plus details of the results you are likely to get - all achievable with only what I would call a typical enthusiast's toolbox.

If you have a Midget/Sprite and are relatively inexperienced at modifying for high performance, then I can almost guarantee that this book will pay for itself from the first significant job you tackle.

David Vizard
California

Essential Information, Using This Book & Toolkit

ESSENTIAL INFORMATION

This book contains information on practical procedures; however, this information is intended only for those with the qualifications, experience, tools and facilities to carry out the work in safety and with appropriately high levels of skill. Be aware that we cannot possibly foresee every possibility of danger in every circumstance. Whenever working on a car or component, remember that your personal safety must **ALWAYS** be your **FIRST** consideration. **The publisher, author, editors and retailer of this book cannot accept any responsibility for personal injury or mechanical damage which results from using this book, even if caused by errors or omissions in the information given. If this disclaimer is unacceptable to you, please return the pristine book to your retailer who will refund the purchase price.**

This book applies to all MG Midgets and Austin-Healey Sprites and, with the exception of the engine, gives every aspect of the car equal prominence so that you'll be able to uprate the whole car. Coverage of internal modifications to the engine is deliberately limited: such coverage would have made this book much, much bigger and an A-series engine tuning 'bible' already exists in the form of David Vizard's well-known book.

Please be aware that changing component specification by modification is likely to void warranties and also to absolve manufacturers from any responsibility in the event of component failure and the consequences of such failure. It is also possible that changing the engine's specification will mean that it no longer complies with exhaust emission control regulations in your state or country - check before you start work. Increasing the engine's power will place additional stress on engine components and on the car's complete driveline: this may reduce service life and increase the frequency of breakdown.

An increase in engine power and therefore performance will mean that your car's braking and suspension systems will need to be kept in perfect condition and uprated as appropriate.

The importance of cleaning a component thoroughly before working on it cannot be overstressed. Always keep your working area and tools as clean as possible. Whatever specialist cleaning fluid you use, be sure to follow - completely - its manufacturer's instructions and if you are using petrol (gasoline) or paraffin (kerosene) to clean parts, take every caution necessary to protect your body and to avoid all risk of fire.

USING THIS BOOK

Throughout this book the text assumes that you, or your contractor, will have an appropriate workshop manual to follow for complete detail on dismantling, reassembly, adjustment procedure, clearances, torque figures, etc. This book's default is the standard manufacturer's specification for your model so, if a procedure is not described, a measurement not given, a torque figure ignored, you can assume that the standard procedure or specification for your car needs to be used.

You'll find it helpful to read the

whole book before you start work or give instructions to your contractor. This is because a modification or change in specification in one area will cause the need for changes in other areas. Get the whole picture so that you can finalize specification and component requirements as far as is possible before any work begins.

As these cars were built to imperial measurements, these take priority in the text.

For American readers, a glossary of automotive terms will be found at the back of the book.

TOOLKIT

If you intend to carry out any or most of the modifications and work detailed in this book, you will need a reasonable set of tools.

Over the years I have built up a modest tool kit which has enabled me to do every job described in this book as well as routine maintenance. I recommend you purchase only high quality tools as they always work out cheaper in the long run. My choice of brand is Snap-On which come with a lifetime guarantee. I've also had very good results from the Britool tools. I think all the fastenings on the Sprite/Midget are AF but, once you start to fit modified parts, you may find they have metric fastenings, so you'll need to adjust your buying to suit.

Many car accessory shops will hire out more expensive and infrequently used items such as hub pullers, coil spring compressors and hydraulic engine cranes.

Here is the list of tools I recommend you have. Although by no means exhaustive, nearly all of the tools will be required at one time or another:

Set of AF combination spanners
Set of AF ring spanners
Set of AF open-ended spanners
Set of metric combination spanners
Spark plug wrench
Brake adjuster spanner - flat type
Sump plug spanner

Set of Allen keys
Set of feeler gauges
Wire brush
Set of plain head screwdrivers
Set of cross head (Philips) screwdrivers
Hacksaw
Pliers
Combination pliers
Grease gun
Torque wrench
Half-inch drive socket set (largest you can afford)
Mole grips
Impact driver
Set of cold chisels
Dot punch
Pop riveter (riveting plier)
Axle stands
Trolley jack
Ball joint splitter (screw thread type)
Crimping pliers (strippers/cutters)
Soldering iron
Circuit tester (12V)
Heavy hammer
Oil filter wrench
Stilsons (pipe wrench)
Electric drill
Strobe light
Jemmy (crow bar)
Medium flat file
Small flat file
Round file
Electrical extension lead
Brake pipe flaring tool - imperial
Decent tool box to keep everything in

Fasteners

Most of the set screws, bolts and nuts on the Midget/Sprite are UNF thread (Unified Normal Fine) and will be plated steel high-tensile grade S with a tensile strength of 50ton ft/in sq and yield stress of 40ton ft/in sq. Engine and transmission fastener strengths vary but are generally much stronger than the foregoing. Keep an eye out for UNC (Unified Normal Course) bolts which are often used to fasten aluminium components: they are stronger than the thread they are tightened into (unlike normal bolts, the thread will strip

out of the aluminium component before the bolt shears).

Stainless steel fasteners, with a tensile strength of 45.3ton ft/in sq and yield stress of 29.1ton ft/in sq., are not as strong as plated steel fasteners.

When using Nylock nuts remember they can only be used once.

For a race car, or any application where weight is at a premium and the application is in an unstressed area, you could use special lightweight fasteners. One such choice would be the aerospace grade aluminium bolt, screw and nut range from AWF. These fastenings are 60 per cent lighter than steel and have a tensile strength of 35 tons. An alternative is fasteners in BT16 titanium which is 45 per cent lighter than steel. This particular grade of titanium is more elastic than other grades and will bend before it snaps, which can be handy. I must emphasise, though, whichever lightweight material you choose, the component should only be used in an unstressed area. A good use for lightweight fasteners, for example, would be retention of the window winder mechanism and possibly the door catch striker plate.

Wire braided hoses

When you look through the Goodridge catalogue for the first time you may be daunted by the unusual terminology used. BGC Motorsport Components prints a simple guide to the Goodridge system which they have kindly allowed to be reproduced here. The Goodridge system is based on UNF (same as JIC) hose ends which are compatible with most things you will find on your Midget/Sprite. Unfortunately, the most likely exceptions will be the very parts you intend to plumb, such as fuel and oil lines, which may use metric, BSP or NPTF threads. The recommended solution to such incompatibility is to use an adapter with UNF fittings at each end of the line.

To work out what bore size hose you require, you'll need to understand the 'dash size' which is used to associate a hose with a thread fitting, as follows -

SPEED**PRO** SERIES

Dash size	JIC thread	BSP thread	NPTF thread
	Guide to Goodridge hose sizing		
-3	3/8 x 24	1/8 x 28	1/8 x 27
-4	7/16 x 20	1/4 x 18	1/4 x 18
-5	1/2 x 20	-	-
-6	9/16 x 18	3/8 x 19	3/8 x 18
-7	5/8 x 18	-	-
-8	3/4 x 16	1/2 x 14	1/2 x 14
-10	7/8 x 14	5/8 x 14	-
-12	1.1/16 x 12	3/4 x 14	3/4 x 14

Chapter 1
Engine

INTRODUCTION

This chapter should provide you with a few useful ideas and, although I don't have direct experience of all of them, I'm satisfied that they are all workable and most are relatively straightforward and practical in application.

TIMING CHAIN CONVERSIONS

When you are building your engine it is well worth the minimal expense of purchasing a Kent Cams vernier cam drive which will give you infinitely variable valve timing adjustment: for further details I refer you to David Vizard's book *Tuning BL's A-Series Engine*. Also in the same book is the Mini Spares toothed-belt drive system.

Unfortunately, during my first engine rebuild, it was not possible to get a suitable cam drivebelt system as, at that time, toothed-belt drive kits did not have an engine breather in the casing. This is no longer a problem as the Mini Spares belt drive kits now have a breather built in. However, on my kit I found that the actual

breather hole was only an 1/8 inch (3mm) diameter so, after checking with Mini Spares that it was okay to do so, I drilled it out to 1/4 inch (6mm) diameter, being careful not to damage the threaded section of the breather. However, I found, as I suspected, that the larger hole breaks

through to the inner part of the cover. This was not a major problem as the hole was sufficiently small to be alloy welded. The larger hole will improve breathing from the crankcase but, since the breather is still quite small, I strongly recommend that if you are using the Mini Spares belt drive

Components of Mini Spares timing belt kit. Note the oil breather outlet at top left of the main casing - this is an important feature missing from some other kits.

Vernier-type cam sprocket. Bolt slots allow precise valve timing to be achieved.

Mini Spares cam drive sprocket featuring offset dowel valve timing adjustment.

conversion you fit a breather outlet to the engine rocker cover too.

Returning to the belt drive kit breather, you can use the short breather pipe supplied with the kit in conjunction with a K&N filter or run a length of hose to an oil catch tank. I used neither of these options, preferring, instead, to fit a Goodridge fitting direct to the casing and run a length of braided steel hose to my catch tank - Hobbsport made up the hose for me.

The original Mini Spares belt drive kit incorporated a vernier adjustment for precise valve timing and this is now one option of two. The other option offers an eccentric dowel method of adjustment. The dowel-type is more fiddly to adjust than the vernier-type, but is more popular for racing applications since the possibility of timing drift, no matter how small, is removed. One of the advantages of this kit is that the removable cover (which is in two halves) allows cam timing changes to be made without having to lift the engine to get the crankshaft pulley off, unlike the standard timing gear and cover. The kit comes complete with a comprehensive set

of fitting instructions.

Finally, a cogged-belt cam drive system gives more accurate valve timing, is capable of standing higher sustained engine speeds and is quieter in operation.

V-BELT SIZES FOR NON-STANDARD DRIVE PULLEYS

Getting the right size V-belt for the car's engine after fitting oversize pulleys may suggest the arduous task of testing various V-belts until the correct one is found, but this need not be the case. Once you have all the non-standard pulleys fitted to the engine, run a piece of rope around the pulleys, marking the rope when you have a complete loop. Measuring this length will give an approximately correct fan belt length. Since there is some adjustment on the dynamo/alternator, it should be possible to get the right size V-belt first time by asking for a particular length from your local accessory shop. An example is the use of the Mini Spares reduced speed dynamo pulley and reduced speed water pump pulley on a 1275cc engine which gives a requirement of around 35.5ins (90cm), the same as the post-1976 Triumph TR7. The 'V' section is also very important, the correct section is 0.38in (9.7mm).

TOOTHED-BELT ANCILLARY DRIVES

In my opinion the ultimate drivebelt system comes from Mini Spares in the shape of a kit consisting of lightweight pulleys in toothed-belt form. The crank pulley is made of steel and will accept the split damper ring from the earlier cars, again available from Mini Spares. It is possible to have water pump drive only, or alternator and water pump drive with a standard size water pump pulley. A larger water pump pulley is also available. The toothed-belt drive is designed to provide a positive drive and helps prevent turned or thrown belts caused by misalignment or very high rpm/engine acceleration.

CAMSHAFTS FOR ROAD ENGINES

Tuning BL's A-series Engine by David Vizard is the authoritative source on cam selection for the A-series engine, but I would like to add that Kent Cams in the UK, A.P.T. and Mini Mania in the USA are companies which can also offer excellent advice on this subject. I recommend either the Kent MD276 or MD286 pattern cams for road use. A third choice, and an alternative to the MD286, is the 286SP: a scatter pattern design. Try to obtain a cam ground from a steel billet, it will last much

Hi-lift rockers come in many forms; these, with rollers, are by Kent Cams.

longer than a normal cam.

Whatever cam you opt for, if you can, use hi-lift (1.5 ratio) rockers (roller type, if possible).

CARBON FIBRE PUSHRODS

Carbon fibre pushrods are usually 5/16in (7.93mm) in diameter, which is larger than the standard pushrod. Therefore, their use will mean that the engine block and the cylinder head pushrod holes will need boring out to take their larger diameter.

Carbon fibre pushrod from Mini Spares.

Mini Spares has tested a carbon fibre pushrod with a diameter of 1/4in (6mm), the same diameter as the standard pushrod. The results showed the carbon fibre rod to be one-and-a-half times stronger and half the weight of the standard item. These standard-sized carbon fibre pushrods can be used as direct replacements for the standard items. Aside from their strength advantages, I strongly recommend carbon fibre pushrods on two counts: 1) a set of these rods is only fractionally more expensive than standard rods; 2) the rods are much quieter in use since carbon fibre does not resonate and also allows the use of smaller valve/rocker clearances. Mini Spares supplies the rods in the UK and Mini Mania retails them in the USA.

CYLINDER HEAD MODIFICATIONS

Unleaded fuel conversion

If you are concerned about the gradual demise of leaded fuel and the likely problems it will cause in running your Midget/Sprite, help is at hand.

It's fairly straightforward to have the cylinder head converted so that your car can safely use unleaded fuel. The main problem with using unleaded petrol is that the absence of lead will cause valve seat recession through wear and valve stem

An example of what a decent modified head should look like. This one is intended for use with unleaded fuel.

and guide wear. Most, if not all, modern engines get around this problem by using harder valves, seats and guides.

The author took his car's cylinder head to Oselli engineering for an unleaded conversion. He went to Oselli as the company originally modified his 1275 cylinder head to a rally big valve specification: with excellent results borne out by both road and dynomometer performance. However, it quickly became apparant that the unleaded conversion comes with a limitation on valve size, especially exhaust. This meant that the exhaust valve size dropped from 1.218 to 1.156in (30.93 to 29.36mm). Of course, you may only require standard size valves, though having a big valve conversion and porting done while 'going unleaded' is cost-effective.

Oselli were sure that, despite the slight reduction in exhaust valve size, in terms of power, the drop would only be around 3 to 4bhp over the original head. A bigger inlet valve could have been used but, in terms of power, might not have produced any gains at all since the extra mixture going into the head could not get out fast enough without a correspondingly larger exhaust valve. Oselli also dropped the CR on the author's car's engine from a high (for a road engine) 11.0-1 to 10.5-1 (though the car will still be run on super unleaded [octane boosted] fuel only.

If you have an existing modified head with big valves, it may be best to sell it to a racer and then buy a stock head for the unleaded fuel modification.

A drawback of the unleaded conver-

Specially machined bronze valve guides are used in this 12G940 head for unleaded fuel use.

sion is that anti-reversionary Rimflo valves are not suitable. This should not be a serious problem if you are using a CV exhaust manifold. The author has not noticed any lack of bottom end flexibility without the Rimflo valves, and suspects the CV manifold is making good any deficit.

Once the high performance unleaded head was installed and the engine run-in, the car was taken to Peter Baldwin to have the engine dyno-tested. Surprisingly, the new engine did not make anything like the horsepower of the old engine, though Peter suspected that this was more likely due to a mis-timed camshaft than a problem with the head. In addition, the engine had valve bounce at just 6000rpm: pretty poor considering a standard engine will not experience valve bounce until some way past 6000rpm. The author's currently awaiting a reply from Oselli concerning this problem, but has received nothing so far . . . The author still believes that the high-performance unleaded conversion should work and has certainly not heard of valve bounce problems on

Arden 8-port head from Mini Spares fitted to Dave Gilbert's road car. Dave is using Amal carbs on this engine. *(Courtesy Dave Gilbert)*

standard unleaded conversions. He plans to take his engine to another engine builder who has experience of unleaded fuel cylinder heads.

Another point to consider while head work is in progress is whether at any time in the future you may use carbon fibre pushrods. These rods usually require larger holes than standard rods, especially if you are using hi-lift rockers, so it's worth having the holes enlarged while other work is being done.

Lastly, while the head is off the car, now is the time, if you are using a blanking sleeve, to plug the bypass outlet in the cylinder head.

Arden 8 port cylinder head
Because this head is designed for the A-series engine it will, of course, fit the Midget/Sprite and Dave Gilbert, a member of the Sprite & Midget Club, was, at the time of writing, fitting one to his car. Dave advises me that it fits just fine with problems limited to moving the odd item and a dynamo to alternator swop. The advantage of this modification is not only the weight saving achieved by switching to

Arden head provides four separate exhaust ports too.

Impressive MK Parts racing Arden-headed engine uses twin sidedraught Weber carbs.

an aluminium head, but also that its superior design allows race-type power output with better than standard road mid-range power and torque. So, you get a really tractable engine that pulls well from

the bottom of the rpm range right up to high rpm with lots of power all the way. It's not cheap but, then again, not expensive when compared to any other modification that will produce this amount of power. Aside from Dave's car, the only other one I know of is the MK Parts car in Germany. The head itself is available from Mini Spares in the UK and Mini Mania in the USA.

Extra head stud & bolt
A modification (applicable to A-series 1275cc engines) that can be done while head work is being carried out is drilling the head and block to take an extra cylinder head stud and bolt. All 1275 Cooper S engines had these extra fittings and if you are using a Cooper S head

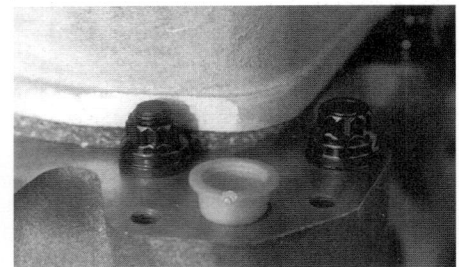

The flange of this rocker cover has been relieved to clear the extra stud. Note the neat competition studs and nuts (available from APT).

you'll already have the holes.

Head & manifold studs
Still on the subject of cylinder head studs and if you are building a high performance engine I recommend you consider using aircraft-quality studs and 12 point nuts from APT in the USA (available through A.P.T. Concessionaires in the UK). These

Competition-quality head retaining stud and nut set from APT.

studs are strong enough to take 190,000psi yet their cost is only around twice that of standard studs and nuts: I consider they are worth every penny. The APT stud kit uses a stud rather than the Cooper S bolt at the front of the block but, like the standard bolt, the stud nut is torqued to only 25lbs (11.34kg). If you have a 1098cc engine you can also use APT studs, but will require only a 9 stud kit. When fitting the studs, ensure that the threads in the block are cleaned out with a square bottom tap. The new studs should be screwed in finger-tight - about 4lbs (1.81kg) of torque without using thread locking compound. The stud holes in the block and head should have a generous chamfer.

APT also has a very neat stainless steel stud and 16 point nut manifold set for the A-series engine. Apart from the fact that it looks really neat, the advantage is that the nuts are very small, a big advantage if you are using a tubular exhaust manifold and non-standard inlet manifold, *i.e.* for a Weber. The advantage is that you can get a socket on a greater number of the nuts and a ring spanner on the remainder. What I did find, though, was that I needed a long 3/8in AF spanner to get decent leverage as my existing 3/8in spanners were on the short side.

ROCKER (VALVE) COVER

There are numerous replacement rocker covers available for the A-series engine, usually made in polished alloy. I have an Oselli cover on mine as that company built the engine. You might wish to fit a similarly smart alloy cover from Motobuild or use an MG Metro item.

If you have a racing set-up with an oil breather from the rocker cover, you'll need a breather take-off point in the cover.

You may wish to replace the stand-ard rocker cover securing nuts with the Mini Spares' T-bar nuts: handy for quick access and nice and shiny, too!

If you have modified your engine's block and head to take the extra stud and bolt (or second stud if using APT studs),

The inside wall of this rocker cover has been relieved to give clearance for hi-lift rockers. The outlet for the oil breather is positioned between rocker pillars.

you'll need to relieve the rocker cover base to ensure the cover fits around the extra stud and nut. This is easily achieved with a file or rotary burr but DO NOT refit the cover without first cleaning thoroughly to remove all aluminium swarf.

ENGINE BACKPLATE

Motobuild or Peter May Engineering can supply an alloy engine backplate which is approximately two-thirds the weight of the standard item: a useful weight saving for the serious racer. In the USA, alloy engine backplates are available from Tom Colby's Speedwell Engineering. Note that backplates are specific to each engine size but that the 1098cc engine backplate can be used on 948cc engines to allow the use of the later ribcase-type gearbox. Note that, when you fit an alloy backplate, you'll need to remove the oil pump cover plate from the old backplate. It's also helpful to mark the bolt hole positions for

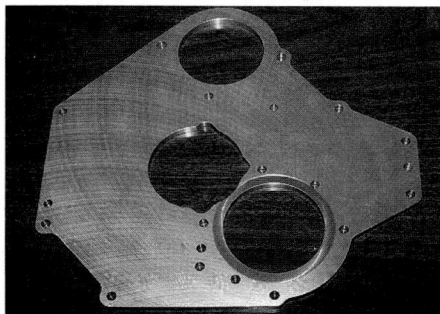

Alloy engine backplate from Speedwell Engineering. *(Courtesy Tom Colby, Speedwell Engineering).*

the alloy plate by using the standard plate as a template. This latter point prevents confusing backplate holes with any of the gearbox to block holes. The respective weights of the standard/alloy plates are: 8lb 8oz/3lb 9oz (3.99kg/1.77kg), the latter representing a weight saving of close to 5lb (2.27kg).

MOUNTINGS & TIE BARS

For the racing or high performance road car, Motobuild can supply competition engine and gearbox mountings made of stiffer rubber than the standard items.

Peter May Engineering fabricates engine/gearbox steady (tie) bars. These bars are rose-jointed (metal spherical joints) and prevent excessive engine movement under hard acceleration and braking.

DYNAMO PULLEY

Owners of early models will have a car equipped with a dynamo and voltage regulator control box. This is quite satisfactory for normal use but with very high revving, high performance engines

Big dynamo pulleys - Longman (left) and Mini Spares.

the dynamo can over-speed which will cause its destruction. The simple solution is to fit the Mini Spares reduced rpm dynamo pulley. Richard Longman also manufactures a similar pulley but of a larger diameter and with a deeper 'V' to prevent the fan belt jumping the pulley at very high rpm. The author has used the Mini Spares item on his car and found it successfully prevented dynamo over-speeding (safe to 7600 engine rpm) and still gives an adequate charge. The

Longman item is better suited to a race applications but its lower gearing would not allow high enough dynamo speeds to charge the battery during normal road use. Use and fitting of either pulley requires removal of the old item and, in either case, a slightly larger than normal fan belt will be required.

Even if your car is fitted with very high wattage headlamps, output from the standard dynamo with large diameter pulley should still be good enough for everyday use, though you'll probably have to adjust the voltage regulator with the aid of a voltage meter. However, if you're going night rallying or have extra lights, the Lucas C40L-type dynamo will provide more output - that is, if it's not already fitted.

CONVERSION FROM DYNAMO TO ALTERNATOR

Having said in an earlier edition of this book that I considered this a job for an experienced auto electrician, I stand corrected by Danny Fenton of the Bristol branch of the MGOC. That said, converting from dynamo to alternator is a little bit more involved if you have an FIA battery master switch. Note that the use of any other type of switch risks damaging the alternator, should the switch be thrown with the engine running (something which is not a problem with a dynamo.

The advantage of using an alternator, as opposed to a dynamo, is that, not only does the alternator have a higher output, but it's produced across a wider range of engine speed, including very low tickover. A dynamo is used in conjunction with a regulator unit, sometimes known as a control box. The alternator also requires a voltage regulator unit but this is built into the alternator itself.

A feature of many alternator conversions is that, for simplicity of wiring, the regulator box is retained in the wiring circuit, though it is no longer being used to regulate current but acts instead as a junction box. Note that a dynamo produces DC current and an alternator

An alternator is easily fitted in place of a dynamo on earlier cars.

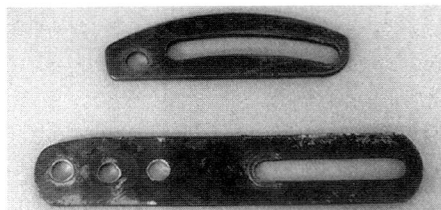

The adjustment strap for the alternator (top) is different to the dynamo version (below). The alternator bracket shown has extra location holes to suit the unusual pulley combination of the author's car.

produces AC current. AC current cannot be used to charge the battery, so it is converted to DC as it leaves the unit by rectifier diodes.

Moving on to the parts required for the conversion, I found that the cheapest way to convert to an alternator was to visit a local breaker's yard to get the requisite brackets and alternator unit. It doesn't really matter if the unit is clapped-out since alternators are generally a service exchange item, so money is not wasted if the unit needs to be exchanged. Minis, Metros and Marinas can all be used as donor vehicles, as can the MGB with a slightly higher output unit. A Lucas 16ACR unit is the model to fit and has an output of 34 amps. However, the 17ACR unit, with an output of 36 amps, will also fit. Note that late 1275cc Midgets and 1500cc Midgets that were equipped with an alternator as standard, sometimes have different outputs quoted. The lower, 16ACR, output should suffice for all applications.

Note that the alternator adjustment bracket may only allow only a negligible amount of movement before fouling on the front engine plate and engine mount-

ing; if so, use the original dynamo-type adjusting bracket. The standard-sized alternator pulley will be okay even on highly-tuned cars (maximum safe alternator rpm is 15,000). The use of a larger pulley increases the cutting-in speed and maximum output speed, but does not reduce maximum output unless 6000 alternator rpm cannot be achieved. If you do want to fit a larger pulley than the standard 2.36 or 2.5in (60.0 or 63.5mm) pulley, the Mini Spares pulley with 3.93in (100mm) diameter will fit the alternator as it does the dynamo. Fitting may require a shim to be made to space the pulley from the fan. I used a phosphor bronze trunnion shim with a notch filed in it to allow fitting. Once the pulley and fan are fitted, it may be necessary to reposition the alternator to ensure the pulley is aligned with the water pump and crankshaft pulleys. The cut-in speed with the Mini Spares pulley is likely to be around 3000rpm but, as previously mentioned, this is not going to be a problem, even with a road car, and the alternator will certainly produce a greater charge than a dynamo with the same pulley.

Remember, when you remove the alternator from the donor vehicle, to cut off a long length of wire - which includes the plug (which can remain in the unit) - from the donor vehicle wiring loom. If you prefer to use new wiring then the alternator plug and all the requisite wires can be purchased from Merv Plastics Ltd. It is a condition of some alternator guarantees that a new plug is used, since resistance in the old plug is claimed to burn out diodes in the new unit Make a note of the make, model and year of the donor car so you can refer to a workshop manual to ascertain colour codes for the connector block. If you re-use the connector block the connectors can be released individually from the block and fresh wire soldered to them. The use of butt connectors to connect the ends of the wires you cut with the connector block to new wires is not recommended because it can cause voltage drop problems.

Remembering that your car must be negative earth, you will need to refer to the workshop manual wiring diagram for the colour codes for your wiring. The wire that was the ignition warning light wire from the dynamo to regulator box and to the light can now run directly from the alternator indicator terminal to the bulb. The two remaining terminals on the alternator both go to the solenoid. The regulator box can be removed. The main feed from the regulator box can now go to the solenoid as well. The wire from the light switch to the regulator box can now can go direct to the fuse box. Note that the other wire from the same terminal on the regulator box (the feed to the ignition switch) must be on the same side as the feed to the fuse box from the solenoid. If you get this terminal feeding across a fuse you'll be drawing the entire electrical load across a single fuse and it will blow, usually when you put the lights on. The earth wire from the control box can be discarded, as can the old dynamo main feed to the control box. In practice, it's not as complex as it sounds but, since there are numerous different looms using different colours, it's not possible to give a colour-by-colour explanation. If you experience problems with fuses, fit an ancillary fuse box. Merv Plastics can offer a variety of types.

Once you have the alternator fitted and running you may need to adjust engine idle speed, especially if the engine is a race tuned unit using a long duration cam. The reason for this is that when any electrical item is switched on (eg a Kenlowe fan) it draws a current which the alternator responds to. On a race-tuned engine which is not making a lot of power at idling speeds, this extra-ancillary power demand can make the difference between the engine idling or stalling. The overall power demand of an alternator is slightly higher than that of the dynamo, drawing some 1.8 to 2.5bhp on full output compared to 1.5bhp of the dynamo.

ENGINE OIL - GENERAL

Engine oil is not only a lubricant but also a coolant, the oil dissipating approximately

ENGINE LUBRICATION OPTIONS

Intended usage	Later type filter head	Oil cooler	Oil thermostat	Sump modifications
Std/mild road	Yes	No	No	No
Medium road	Yes	Yes (10 row)	Yes	No
Fast road/ Mild competition	Yes	Yes (13 row)	Yes	As required
All-out competition	Yes	Yes (16 row plus)	No	As required

10-15 per cent of engine heat. A high engine temperature will produce a high oil temperature. As the engine oil becomes hotter it gets thinner and, consequently, less protective so, obviously, oil choice and temperature are important. A cheap oil will break down more quickly at high temperatures and the very best (and most expensive) protection will come from a good synthetic oil.

I have used Mobil One synthetic for many years as it has a reputation for being among the best. However, note that, generally, Mobil One is not suitable for running-in engines and the A-series - probably the 1500 Midget too - are no exceptions as bore glazing can result.

Mobil One will withstand temperatures of up to 320 degrees F (150 degrees C). However, I understand that oil temperatures of between 203 and 230 degrees F (95 and 110 degrees C) will produce the best engine power output. To ensure your engine is running at the optimum oil temperature you'll need to monitor it by fitting an oil temperature gauge. Once fitted, you'll be able to determine the cooling requirements of your engine.

Engine lubrication options

The accompanying chart shows the author's engine lubrication system recommendations for the Midget/Sprite in various states of tune.

Later-type filter head allows use of disposable cannister filters. Note braided oil cooler hose.

Later type filter head

A worthwhile modification, which can be made whilst the engine is having an oil and filter change, is the fitting of a filter head bracket that will take the later canister-type of oil filter. The new bracket can be obtained from Mini Spares Ltd in the UK and Mini Mania in the USA.

The old filter head bracket and case are unbolted from the block and discarded, then the new bracket is bolted to the block using a new gasket. The job is as simple as that! The canister-type oil filter can now be fitted.

The main advantage in having a canister filter is that it can be filled with oil before fitting, which enables engine oil pressure to be restored more quickly after an oil change. A canister filter is also a lot less messy to replace than the element type.

An alternative is to blank off the filter outlet on the block and fit a completely remote filter set-up.

When you buy a filter, I recommend you use a genuine Unipart item (GFE 1) as cheaper copies have been known to seriously degrade engine oil pressure because of poor internal design. An additional advantage of the Unipart filter is that it is the only one I know of to incorporate a magnetic trap.

Oil cooler

All 1500 Midgets will benefit from the fitting of an oil cooler since they tend to run on the hot side, even with moderate driving. A-series-engined cars are not likely to need an oil cooler unless they have a modified engine or run in a hot climate. For any car, if you have established the temperature range of the engine oil (by fitting a gauge) and finding it hotter than 230 degrees F (110 degrees C), you should fit an oil cooler as a priority.

An oil cooler is a radiator that exchanges heat between the oil and the air: the flow of air through the matrix of the cooler unit cools the oil passing through it. The fitting of a cooler also increases the total oil capacity of the system by approximately: 10 row - 0.23ltr, 13 row - 0.30ltr, 16 row - 0.37ltr (these figures apply only to the common, long thin type of cooler but, should capacity be an important consideration, check with the manufacturer the amount the cooler and its piping holds).

Before fitting a cooler, the optimum size must be determined and this be done by deciding how much a temperature reduction is needed. Ten rows is normal for a well-tuned road car; thirteen for high performance road cars. For competition or very high performance road cars (120bhp+), 16 rows or larger may be necessary to keep the oil temperature around the 212 degrees F (100 degrees C) mark. To prevent over-cooling of the oil in cold weather, and to assist initial warm-up, an oil thermostat can be fitted to road cars.

Oil coolers can be bought in kit form, which usually comprises a cooler unit, all pipes and fittings and mounting brackets. However, some kit brackets are not particularly neat and an alternative is to

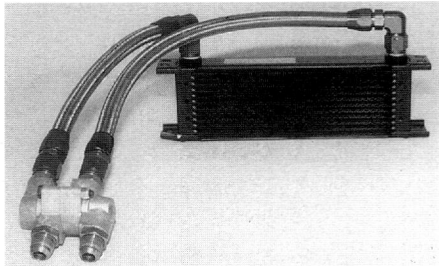
Setrab oil cooler with steel braided hoses and a thermostatatic flow valve.

buy a bracket separately from a good MG parts supplier under part number AHA 8386, or fabricate your own.

The cooler should be positioned where it will receive good airflow. In my experience, bolting it in front of the radiator but behind the grille is near ideal as the cooler receives maximum airflow whilst the radiator remains largely unaffected, despite what some people might tell you. This position also keeps the oil cooler reasonably well protected from flying stones and other road debris - remember a punctured oil cooler or pipe could result in sudden and complete loss of oil pressure.

To fit an oil cooler, first remove the copper pipe connecting the engine block to the oil filter head and discard it. There are fittings in the kit to screw into the two outlets. From the fittings, the oil pipe runs to the cooler unit: it does not matter which pipe runs to which end of the cooler as oil flow in the cooler can be left-to-right or *vice versa*.

It's possible to use Goodridge-manufactured stainless steel braided hose and anodised alloy fittings in preference to the rubber hose supplied with a kit. The advantage lies in the stronger construction and higher chafing resistance of the steel braiding and, as a bonus, it looks extremely businesslike. I have also found that the normal pressure drop of 10psi (69kpa) or so on fitting a cooler using rubber pipes is practically eliminated when using Goodridge steel braided hose.

Oil cooler thermostat

The only drawback with fitting an oil

cooler is that in the winter it's possible to overcool the oil. To prevent this and to speed oil warm-up time, an oil thermostat will be required.

I have used Goodridge/Setrab oil thermostats for some years now, and strongly recommend them. Be advised, however, that some thermostats can be restrictive.

An oil thermostat is easy to fit to your oil cooler piping circuit. The two hoses leading to and from the oil cooler are cut in two, leaving you with four hose ends. Locate the union stub on the thermostat marked "Inlet," to which the hose leading from the filter head must be connected; the hose from the engine block should be connected to the union stub marked "Outlet." This allows oil to flow from the engine through the thermostat and back to the engine when the thermostat is in the closed position. The two remaining pipes from the thermostat lead to, and from, the cooler unit. When the oil reaches an operating temperature of around 172 degrees F (78 degrees C), the thermostat will open to divert oil flow all the way to the cooler before it returns to the engine.

Oil thermostats come in several sizes but a large unit is recommended. Available from Goodridge, they can have either push-fit with hose clip or, tidier, threaded

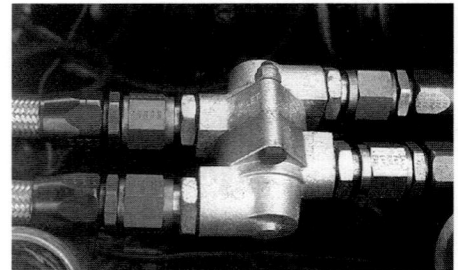
Close-up of thermostatic flow valve in oil cooler lines.

hose fitting connections. Whichever installation you choose, before you drive off in your car, check for oil leaks before and after, allowing time for the oil to reach working temperature. I have heard that the push-fit and hose clamp connections can come off, though I didn't have any problems during the time I used this type of fitting. The completed job, if done correctly, should prevent damage to your engine through overheating, a lesson the author learnt the hard way by virtue of a scuffed piston and blued piston pin bush (small end).

Oil breather(s) and oil catch tank

When tuned to produce extra power, A-series engines benefit from extra crankcase ventilation; the breather pipe(s) going into an oil catch tank. You may want to fit an oil catch tank anyway, whether your car's engine is A-series or 1500, since it is usually a mandatory requirement for motorsports.

The existing breather will be on either the engine side cover or the timing cover (1275 A-series) and may, or may not, feed back into the engine induction system. A supplementary breather can be run from the engine rocker cover. On the standard steel rocker cover you'll need to drill a hole of 1.5in (127mm) to which a short length of pipe can be welded. The same method *can* be used on alloy rocker covers but I prefer drilling and tapping the hole and using a Goodridge threaded fitting. A point to note is that the hole should be marked whilst the rocker cover is fitted to the engine but drilled whilst removed from the engine. The main reason for this is that marking the hole while the rocker cover is off the car can lead to bad positioning of the breather hose; I found this out the hard way and have minimal clearance between oil breather connection and servo hose. The second reason is that drilling the hole with

Rocker cover breather fitted to the author's car. It would be better if the union were higher up the rocker cover.

An oil breather hole in the cover is essential when fitting a belt drive conversion to the Sprite/Midget. This is the Mini Spares cover.

Here, timing cover breather hole is vented via Goodridge braided hose (installation by HobbSport).

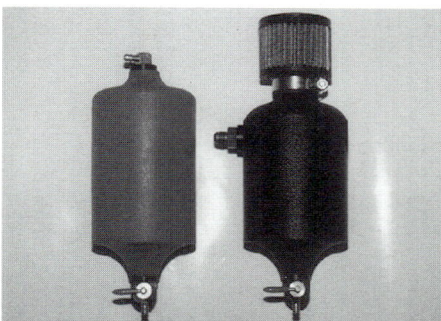

Oil (with breather) and water catch tanks both with drain taps: by Speedwell Engineering. (Courtesy Tom Colby, Speedwell Engineering).

the rocker cover on the engine can introduce swarf to the engine.

Any rubber or braided steel hose of the appropriate diameter can be used to complete the installation by feeding into the oil catch tank. An interesting point is that since the engine vents oil when it breathes, especially at high rpm, some racers feed the standard crankcase breather, via a pipe, into the rocker cover rather than run a long length of pipe to the oil catch tank. This is a neat idea which

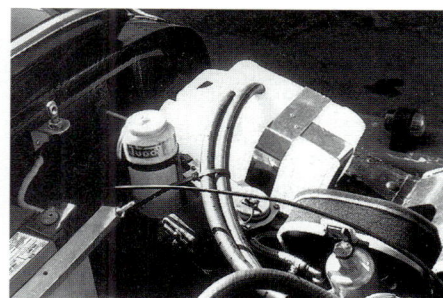

Neat home-made catch tank on Geoff Hale's car.

has some merit and the only drawback is that it will require extra work to enable the hose to fit onto the rocker cover.

Most racers use an old plastic milk or squash container as an oil catch tank though it is possible to use a purpose-designed item made of aluminium available from Tom Colby's Speedwell Engineering. Whatever type of tank you use, it will have to be secured to the car by bracketing which you can make yourself.

Oil sump

There are modifications that can be made to the sump: BL Special Tuning did make an extra deep sump and modified pick-up pipe for the Midget/Sprite at one time under part numbers C-AJJ 3324 and C-AJJ 3323 respectively. The only real advantage, I'm told by Rae Davis of Motobuild, is that the extra capacity allows for some oil consumption during a long rally or race. If you do a lot of long distance rallying and racing, you might try to get hold of this sump and pick-up, though it would probably be simpler to cut two sumps in half - one cut near the bottom and the other near the top - and then weld the two sections together to provide the required increased depth: you'll also need to extend the pick-up pipe.

The Marina sump will fit the Midget/Sprite but has an extra plate on the underside which needs to be cut off. If you find oil surge a problem, either during cornering or under severe braking, then Peter May Engineering Ltd can supply you with an exchange baffled sump or baffle your existing sump.

A further sump modification is to fit

Inside the Speedwell Engineering super competition sump with removable windage tray (arrowed) and modified oil pick-up. *(Courtesy Tom Colby, Speedwell Engineering).*

Fitted Speedwell Engineering big sump conversion, complete with breather. *(Courtesy Tom Colby, Speedwell Engineering).*

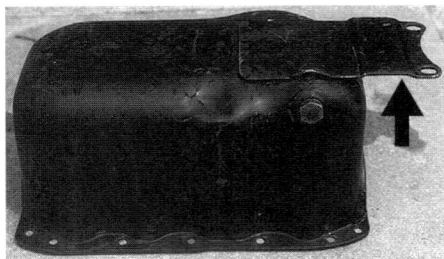

Marina sump. The plate (arrowed) will need to be sawn off.

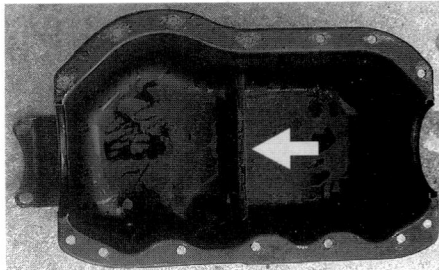

Inside Marina sump. Single baffle plate (arrowed) is insufficient to prevent oil surge on competition or fast road cars.

forward under hard braking and is baffled.

Mini Mania has recently introduced a baffled sump which, from photos I have seen, looks like it has a built-in winding tray.

Whichever sump you use, you'll find a Speedograph sump drain plug to fit it. The advantage of this sump plug is that it has a magnet in the centre which will collect the fine metal deposited in the oil as a result of engine wear. The part number is: DP3 for A-series engines. In the USA, you can obtain a magnetic sump plug from The Winners' Circle.

If you fitted an oil temperature gauge to your car which used a sump sender unit located by a union, and are now fitting either a Peter May baffled sump or a

Standard and magnetic insert (right) oil sump plugs.

Union for sump oil temperature sender.

Speedwell competition sump, you'll need to fit an oil temperature gauge sender union to the replacement sump. It is preferable to purchase and fit a new union rather than try (possibly without success) to remove the one in the old sump. The part number for the union is SIB 733 and it is available from any Lucas parts dealer.

If you intend to rally your Midget/Sprite then it's possible to have a sump guard fabricated by a local metal worker, as did my friend, Phil Bollen who, when he tested it, found it worked. However, remember that the sump guard should be designed to minimise the loss of airflow through the engine bay and around the sump.

Dry sump kit

Huffaker Racing makes a dry sump kit for the A-series engine which is available from Mini Mania. The kit includes a special baffled oil sump with fittings, oil pump mounts, oil pump blanking plate, special fitting to replace filter block, toothed-belts, pulleys and scavenge lines. The kit does not appear to include the oil reservoir which would, of course, be required.

There are two advantages to 'dry sumping', one of which is to allow the engine to sit lower in the car without having problems with sump to ground clearance. On the Midget/Sprite, this benefit doesn't really exist since the chassis H-frame is practically at the same level as the sump. Secondly, there are no problems with oil surge in a dry sump system. The third benefit of dry sumping is some reduction in oil windage power losses. 'Some' is the operative word here; the rpm and power the engine produces need to be of high order for any worthwhile gain to be made. For instance, if you have a really high revving, high horsepower 1275 engine, then the gain may justify the time and expense of dry sumping. Only the hottest race engines will benefit from dry sumping.

ENGINE SWOPS

You can fit the 1275cc A-series engine into cars which had the 948cc or 1098cc

an oil separator breather in the sump to improve crankcase ventilation, venting the outlet to an oil catch tank.

Another possibility is to use the competition oil sump retailed by Tom Colby's Speedwell Engineering in the USA. This is a large capacity sump, which holds approximately 1 quart (1 litre) more than a standard sump. It is also fitted with a hinged trapdoor to prevent oil flowing

A-series engines with very few problems; however, always transfer the larger engine's gearbox and cooling system at the same time. If you use a 1275cc engine from a Morris Marina, you'll need to retain your Midget/Sprite engine backplate and, if possible, the crankshaft and flywheel, etc, too. Fitting the Triumph 1500cc engine into cars which were equipped originally with an A-series engine is a very difficult conversion and you can get just as much power from a tuned A-series unit.

Using the Morris Marina block in the Midget/Sprite

For any owner who needs a new engine for his Midget/Sprite (not 1500) there is a cheap alternative to buying a Midget/Sprite engine (assuming you could find one anyway) and that is to use the British Leyland in-line version of the A-series engine that was fitted to the Morris Marina, or A-Plus in-line engine fitted to the Morris Ital (facelifted Marina).

These engines - or, to be more precise - engine blocks are a popular choice amongst racers. At first sight, it appears they are similar in every respect other than that the oil filter is positioned in a different place. However, in reality, quite a few parts need to be swopped over before the 'Marina' engine can be used.

Firstly, the rear engine backplate from the Midget/Sprite has to be used as the hole configuration on the Marina backplate is for the Marina gearbox, which is quite a bit different to the Midget/Sprite box. Likewise, the front engine plate is different and so is the timing gear case. The Marina front engine plate cannot be used, even with the Marina timing gear case, as it has no provision for engine mountings. The Marina sump has a metal plate spot-welded to it which has to be sawn off, flush with the end of the sump, before it can be used in the Midget/Sprite. Alternatively, the Midget/Sprite sump can be employed.

The engine internals are also different; the Marina block can be used with either a Marina or Midget/Sprite crank. However, each crank can only be used with its respective rods (i.e. Marina crank

and rods or Midget/Sprite crank and rods. Having said this, the Marina crank is common in size to the later A-Plus transverse engine and will take any of the other big bore block rods except for the Cooper S-type. For detailed advice on cranks and rods, I refer you to Tuning BL's A-Series Engines by David Vizard. The Midget/Sprite flywheel can be used with a Marina crank but needs modifying to fit the A-Plus block. The distributor is also different on the A-Plus block.

Because the oil filter is housed differently on the Marina block, the only way an oil cooler can be plumbed-in is by the use of a sandwich plate (available from Hobbsport Ltd or Goodridge UK Ltd). Both the Marina and Ital blocks have provision for an engine-driven fuel pump; the opening can be blanked off with a fabricated plate. However, a better use for this feature is as an engine breather (run a length of hose from an appropriately fabricated cover plate to an oil catch tank). Alternatively, if the engine is turbocharged, you can use this orifice for the turbo-charger oil drain.

When it comes to a cylinder head for a Marina block, there is a wide choice: the

most popular casting is the 12G940 item which can be modified in accordance with the instructions given in Tuning BL's A-Series Engine by David Vizard. Both Marina and Ital 1300 A-Plus heads have casting number 12G970, the standard 9 stud fitting which will benefit from the head fastening modification described elsewhere in this chapter. The Metro Turbo head is also a good one to use as it has larger waterways than earlier heads.

Fiat twin cam engine swop

There are a few Morris Minors which have been converted to take the Fiat 2000cc twin cam engine and, on the face of it, this seems like an ideal conversion. However, although the Fiat engine fits just fine in the Minor, the main problem with putting this engine into the Midget/Sprite is the fact that it's taller than the standard engine and therefore requires a considerable bonnet bulge. Fitting will entail cutting back the front bulkhead and floor section crossmember and then reinforcing everything to take the new gearbox mountings, etc. Other problems include locating the alternator, re-siting the battery and fitting a remote filter.

Fiat twin cam engine fitted in Richard Minns' Midget. Note height of engine and clearance modification to bonnet. (Courtesy Richard Minns).

Rover K-series engine in a Sprite as fitted by Brian Archer.

I know of two Midgets with this conversion, both of which used the 1500cc shell which is larger at the front. This is an interesting alternative to tuning a 1275cc engine but probably more expensive and certainly more work. For a 1500cc-engined car it does have its merits, although I don't recommend it as there are much better options available.

Rover K series engine swop

An engine swop which is still breaking new ground is conversion to the Rover K-series engine. This is an interesting choice for an engine swop and, in my opinion, is likely to provide stiff competition to a KAD, Jack Knight or Mini Spares (Arden) cylinder head conversion in terms of power increase. I think it could also prove to be equally stiff competition to a forced induction A-series unit, and a blown K-series engine would be pretty neat.

My main reason for favouring the K-series engine is its all-alloy construction, which makes it a lot lighter than the A-series unit. Remember, reducing weight has the same effect on performance as increasing power. Also, by reducing weight the overall handling of the car will improve. The K-series engine is also capable of being tuned to a power output for road use that nearly all racing A-series engines would struggle to attain: I have heard claims of a 'chipped' injection 1.4 unit producing 125bhp; a warm road engine on Webers should produce 130bhp. However, I would expect the ultimate road specification engine to exceed these figures.

The engine is strong and can be built to run up to 8000rpm. Kent Cams has a range of cams suitable for road, rally and race uses. Two important facts to consider about this engine are that it is slightly taller

than the A-series (but not the 1500cc engine), and the heater box and battery will have to be moved as the distributor sits at the back of the engine and protrudes a long way.

Even more exciting than a modified 1.4 K-series engine are the two 1.8 K-series engines used in the MGF. This engine is a bored and stroked 1.4 unit with a few crucial differences. To get the extra bore the 1.8 engine uses a special cylinder liner that is wet at the top and dry at the bottom, unlike the 1.4 cylinder liner which is totally wet. (MG call this the "damp liner"!). The stroking is achieved by a special crank with some slight changes to the main bearing ladder casting to enable sufficient clearance for the larger crankshaft. So, dimensionally, the 1.8 engine is the same as the 1.4 which means it will fit the Midget/Sprite.

Things get even better with the other version of the 1.8 engine which has a sophisticated variable valve control (VVC) cylinder head which MG claims is a generation ahead of the Honda VTEC system.

Having driven a Honda CRX with a VTEC engine, the author is looking forward to evaluating MG's claims. Certainly, the Honda VTEC is an amazing engine that that can be run up to the red line of 8000rpm at pretty much every gearchange. The 1.8 VVC K-series promises 145PS straight out of the box, so imagine having that under the bonnet of your Midget/Sprite!

The company dealing with this conversion at the present time is Frontline Spridget Ltd. - talk to Tim Fenna. If you do opt for this conversion, you'll need a Ford-based gearbox to go with it.

Chapter 2
Fuel System

WEBER (& DELLORTO) SIDEDRAUGHT CARBURETTORS

Before purchasing your expensive Weber or Dellorto carburettor, manifold and filter you need to ask the question - does the engine in my car have the power potential to merit a twin choke fixed jet carburettor? The answer is almost certainly yes, because, if you've started down the performance tuning path and you've got a head and exhaust system capable of flowing more gas, you'll need bigger carburettors. If you are going to have to fit something which will flow more air than the standard twin 1.25 inch SUs, you might as well fit a Weber or Dellorto as a pair of bigger SUs. Finally, on a purely aesthetic level, a sidedraught Weber or Dellorto simply looks the business.

Having decided that your engine *does* need a twin choke sidedraught carburettor, the next question is what size? The 40mm and 45mm are the most common sizes of Weber and Dellorto, although other sizes are available. However, for Midgets and Sprites of up to 1098cc, a single 40 is big enough and

Paul Rodman's 1500cc Midget with twin Weber DCOEs on Triumphtune intake manifold. Note filter cases, less filters and outer covers. *(Courtesy Paul Rodman).*

1275cc engines usually need a single 45 DCOE.

Because the author's own experience is of the Weber DCOE carburettor, the Dellorto DHLA is not described in the fitting and tuning procedures in this book. However, Webers and Dellortos are virtually identical in size and are equally tuneable so, if you want more information (on Dellortos in particular), there is a companion book in the SpeedPro series called *How To Build & Power Tune Weber DCOE & Dellorto DHLA Carburettors* which will give you all the information you need on carburettor size and calibration to your suit your particular application.

Owners of 1500cc Midgets will be interested to know that Janspeed is one of the few companies that can supply a twin 40 DCOE Weber kit for the 1500 engine. If you're fitting Webers to a 1500, I'm told that clearance is very limited and the carburettors will almost touch the brake/clutch pedal box.

Fitting Weber DCOE

Depending on the manifold used, the DCOE Weber can take up considerably more room than an SU set-up. This can make it necessary to modify the inner wing panel or even bonnet to allow room for the DCOE. If room is tight, you can always use a pair of short ram pipes and a shallow K&N filter (don't try to get by without either item). My experience with an Oselli

Weber 45 DCOE from author's 1312cc engine, fitted with 40mm chokes and utilising Goodridge braided metal fuel hose. Note servo take-off on manifold.

Janspeed 40/45 Weber DCOE carb and manifold conversion for Midget/Sprite with neat progressive throttle linkage. *(Courtesy Janspeed).*

inlet manifold is that it is only the 7 x 4.5 x 3.25 inch (178 x 115 x 83mm) and 9 x 5.25 x 3.25 inch (229 x 133 x 83mm) filters that will require the special body modifications.

The standard heat shields and breathers fitted to your SU set-up will not normally fit a Weber DCOE and manifold. The breather pipe that ran from the inlet manifold to the timing case can be removed and a small K&N filter fitted to the timing cover in its place. If you are retaining the existing distributor with vacuum advance you are unlikely to find a union for the vacuum pipe on a Weber manifold. However, if you are tuning the engine to the extent of fitting a Weber DCOE, you should already have considered fitting a distributor with a modified advance curve which will almost certainly not have vacuum advance.

Before fitting the manifold consider whether the car will require provision for a servo vacuum take-off, distributor vacuum take-off or even a vacuum gauge take-off. The reason for this is that suitable unions involve drilling and tapping the manifold, obviously a job that should only be done with the manifold off the engine.

When fitting the manifold, be sure to use a large bore competition manifold gasket if the cylinder head ports and the inlet manifold have been enlarged. Large bore manifold gaskets are available from Mini Spares Ltd. in the UK and Mini Mania in the USA.

The DCOE must be mounted on rubber O-ring-type gaskets. These reduce fuel foaming caused by engine vibrations transmitted to the carburettor. Various companies market mountings that get the job done.

Weber throttle linkage

A throttle linkage may be included if you buy a Weber as part of a dedicated packaged kit. Alternatively, there are lots of throttle linkages for the Weber/A-series combination but many of these will not fit without bonnet (hood) modifications, so check before purchase. Additionally, some linkages are only suited to the transverse installation of the A-series engine (Mini,

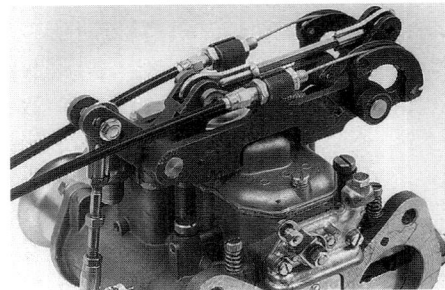

Janspeed twin cable Weber DCOE throttle linkage. *(Courtesy Janspeed).*

Janspeed single cable Weber DCOE linkage modified to clear bonnet line.

Metro, etc.), though these might well fit left-hand drive cars. One linkage that definitely does fit is the Janspeed item which I use on my own car, initially with a bonnet bulge but later modified by Carcraft to fit without a bulge.

Other Janspeed linkages are designed to fit with manifolds that have less or no height restrictions when used on the Sprite/Midget and present no special fitting problems. Janspeed linkages are available from Seven Enterprises in the USA.

Weber operating principles & calibration

Once you've fitted the carburettor it will need to be provisionally calibrated. I say 'provisionally' as the optimum settings can only be ascertained by running the engine on a dynamometer, on which subject there is more later. The accompanying charts show basic Weber calibration for three A-series engine sizes and the 1500cc unit and should be sufficient to get the car running.

Despite its complexity the DCOE is not difficult to calibrate, though on first

Single Weber basic settings for 948cc engines

	Standard 40 DCOE	Road modified 40 DCOE	Race modified 40 DCOE
Aux vent	4.5	4.5	4.5
Choke	31	32	33
Main jet	115	130	140
Air corr	150	155	180
Em tube	F2 or F9	F16	F16
Pump jet	35	40	40
Idle	45F9	50F2	50F2

Single Weber basic settings for 1098cc engine

	Standard 40 DCOE	Road modified 40 DCOE	Race modified 40 DCOE
Aux vent	4.5	4.5	4.5
Choke	32	33	34
Main jet	120	135	145
Air corr	150	155	180
Em tube	F2 or F9	F16	F16
Pump jet	35	40	45
Idle	45F9	50F2	50F2

Single Weber basic settings for 1275cc engine

	Standard 45 DCOE	Road modified 45 DCOE	Race modified 45 DCOE
Aux vent	3.5	3.5	35
Choke	34	36	38
Main jet	145	165	175
Air corr	175	175	175
Em tube	F2	F2	F2
Pump jet	45	50	55
Idle	50F9	55F9	55F2

Twin Weber basic settings for 1500cc engine

	Standard 40 DCOE	Road modified 40 DCOE	Race modified 40 DCOE
Aux vent	4.5	4.5	4.5
Choke	28	31	33
Main jet	115	125	130
Air corr	170	170	170
Em tube	F16	F16	F16
Pump jet	35	35	40
Idle	45F9	45F9	50F9

impression can seem daunting. However, before you work on the DCOE you might find the following rudimentary explanation of its workings helpful.

Think of the DCOE as two identical carburettors in one body, because that is what it is. Don't confuse the DCOE with other Weber carburettors that have one choke larger than the other; these are so-called 'progressive' Webers. The DCOE is a synchronized carburettor with both throttle plates opening equally at once.

Like any carburettor, the DCOE mixes fuel with air to form a mixture which is drawn in by the descending piston and then burnt in the closed combustion chamber. However, unlike most original equipment carburettors, the DCOE can be calibrated in very small increments in order to achieve the optimum fuel/air mixture for your engine.

Once air has flowed through the ram pipe it enters the two venturis in each choke tube, starting with the auxiliary venturi where the airflow draws fuel mixture from the main jetways. The auxiliary venturi acts like a signal amplifier because the end is positioned at the point of greatest depression in the main venturi. (Sometimes the auxiliary venturi is known as the 'booster venturi'). The air (and now fuel mix) having passed through the auxiliary venturi, passes through the main venturi where it speeds up due to the aerodynamic shape narrowing of the venturi.

Because fuel is drawn into the airstream by partial vacuum in the venturi, once the fuel flow starts it is difficult to control the flow precisely, and this can cause excessive enrichment of the mixture. With fixed jet carburettors such as the DCOE, this problem is overcome in part by having a main jet system with air correction, which is controlled by a jet that allows air to be drawn in with the fuel and, in a similar manner, by the holes in the emulsion tube.

The main and air correction jets are supplemented by the idle jet system which provides appropriate fuel air mixture at engine idle and low speeds. The main system provides fuel air mixture at larger throttle openings and stays in use all the way to full throttle; more of this later.

To improve performance the main fuel supply system is supplemented by an accelerator jet known as the pump jet. The pump jet shoots liquid fuel into the venturis when the throttle is suddenly opened and also streams extra fuel when

The carburettor jet inspection cover on Eric Grundy's racing Midget is lockwired, and with good reason: the author can testify that these covers do come undone. *(Courtesy E. Grundy).*

Whichever manifolds you use, the small (3/8in spanner fit) stainless steel studs and nuts from APT make fitting and removal a lot easier and faster, especially when used in conjunction with a non-standard exhaust manifold.

the throttle is in the fully open position.

Now that you understand the basic principles of how the DCOE works, a more detailed look at the three mixture controlling areas completes the overall explanation.

The idle system is regulated by one component but this comes in a combination of jet hole size and emulsion hole size (emulsion is the mixing of the fuel with the air to form fuel droplets suspended in air). So this single component has two calibratible properties. Small adjustments to the idle system can be made with the idle mixture adjustment screw. Remember that, for the main part, when the idle system is working the main system is not and vice versa so they work independently of each other. The only - and obvious - exception to this is the overlap period when one system is finishing and the other starting.

The main system is similar to the idle system but has three calibratible components which, unlike the idle system, are separately removable. The three parts are the main jet, emulsion tube and air corrector which fit together within a holder to form one component.

The accelerator pump jet is calibrated by a jet, spring and a spill-back (or bleed) valve but largely by jet. For most, if not all, A-series applications the spill-back valve in the float bowl will be closed. Whilst the primary function of the accelerator pump is to supply a jet of fuel for sudden acceleration, it can affect the main mixture strength as petrol is drawn from it at high

engine speeds but, as explained above, for the A-series engine the spill-back valve is most likely to be closed. However, on 1500 Midget applications this is not likely to be the case.

Earlier I mentioned the venturis which come in different sizes to suit the airflow and power demand of the engine. Changes in venturi size will need to be balanced by mixture strength changes to keep the air/fuel ratio correct.

Before starting the engine for the first time with the DCOE fitted, note the sizes of all calibrated components, like jets, and keep a record every time you make a calibration change.

Using a dyno IS THE ONLY WAY you can set final calibration to give optimum performance. Choose a dyno operator who is familiar with the DCOE and holds stocks of jets, etc. The first time the author took his car to a dyno with what he thought was a well set-up DCOE carburettor, he still saw a gain of 10bhp on changing main jets! Remember, there is no such thing as a preset DCOE carburettor and jetting recommendations in any book will be just that - recommendations.

Weber air filters & ram pipes

It's strongly recommend that you use K&N filters; in the author's experience their filtering properties are excellent and without detriment to airflow.

There is usually a K&N air filter that will accommodate your chosen ram pipes within the filter body (the standard Weber rams are too long for a Midget/Sprite installation). K&N makes a variety of rams

Weber components
The following is a complete list of the components in a Weber DCOE carburettor: always use the component reference numbers quoted when ordering parts. See the accompanying illustration to match names and key numbers to components.

1	Jets inspection cover	32376.003
2	Screw securing carburettor cover	64700.001
2A	Well-bottom cover screw	64700.001
3	Gasket for jets inspection cover	41550.002†
4	Normal washer	55510.034
4A	Normal washer	55510.034
5	Carburettor cover	31734.025
6	Gasket for carburettor cover	41715.001†
7	Emulsion tube holder	52580.001
8	Air corrector jet	77401*
9	Idling jet-holder	52585.006
10	Emulsion tube	61450*
11	Idle jet	74819*
12	Main jet	73401*
13	Plate for carburettor bowl	52130.003
14	Choke	72116*
15	Auxiliary venturi	69602.*
16	Dust cover	41570.001†
17	Spring	47600.063
18	Spring retaining cover	58000.007
19	Throttle control lever	45034.044**
20	Throttle adjusting spring	47600.007
21	Throttle adjusting screw	64590.002
22	Auxiliary venturi fixing screw	64840.003
22A	Choke fixing screw	64840.003
23	Spring washer	55525.002
23A	Spring washer	55525.002
24	Carburettor anchor nut	34705.004
24A	Nut for air intake	34705.004
25	Retaining plate for air intake	52150.004
26	Stud bolt	64955.104
27	Lock washer	55520.004
28	Hexagonal nut	34710.003
29	Gasket for cap	41640.001†
30	Cap for bottom of bowl	32374.008
31	Carburettor body	N/A
32	Spring anchor plate	52210.006
33	Spindle return spring	47605.012†
34	Lever fixing pin	58445.001
35	Pump control lever	45082.005
36	Stud bolt	64955.007
37	Stud bolt	64955.101
38	Ball bearing	32650.001
39	Throttle securing screw	64570.006
40	Throttle valve	64005.069
41	Throttle spindle	10005.426
42	Starter control securing screw	64700.004
43	Normal washer	55510.038
44	Cap securing screw	64570.009
45	Cap for pump opening	52135.002
46	Gasket for cap	41640.021†
47	Starter control assembly	32556.002
48	Starter control lever, complete with:	45027.030
49	Nut for screw	34720.002
50	Starter lever	45025.029
51	Cable securing screw	64800.002
52	Lever securing nut	34715.010
53	Lever return spring	47610.006
54	Cover for sheath support	32556.001
55	Starter shaft	10085.003
56	Strainer	37000.016
57	Sheath securing screw	64605.017
58	Shim washer	55555.010
59	Starter valve	64330.003
60	Spring for starter valve	47600.005
61	Spring retainer & guide	12775.004
62	Spring washer	10140.012
63	Spring retaining plate	52140/004
64	Pump control rod	10410.015
65	Spring for plunger	47600.064
66	Pump plunger	58602.003

67	Spring for idling adjustment screw	47600.007†
68	Idling adjustment screw	64750.001†
69	Air intake horn	52840.001
70	Screw for progression holes inspection	61015.002
71	Gasket for pump jet	41535.021†
72	Pump jet	76801*
73	Seal	41565.009†
74	Screw plug	61015.008
75	Intake & discharge valve	79701
76	Starter jet	75606
77	Float	41030.00
78	Fulcrum pin	52000.001
79	Ball for valve	58300.001
80	Stuffing ball	52730.001
81	Retaining screw for stuffing ball	61015.006
82	Gasket for needle valve	41535.015
83	Needle valve	79401*
84	Gasket for union	41530.031†
85	Spherical union	10354.001
86	Gasket for union	41530.024†
87	Screw plug for union	12715.008
88	Strainer	37022.002
89	Gasket for filter plug	41530.024†
90	Filter inspection plug	61002.010

*Calibrated parts
†Parts supplied in service kit   93.0015.05
**Varying with application

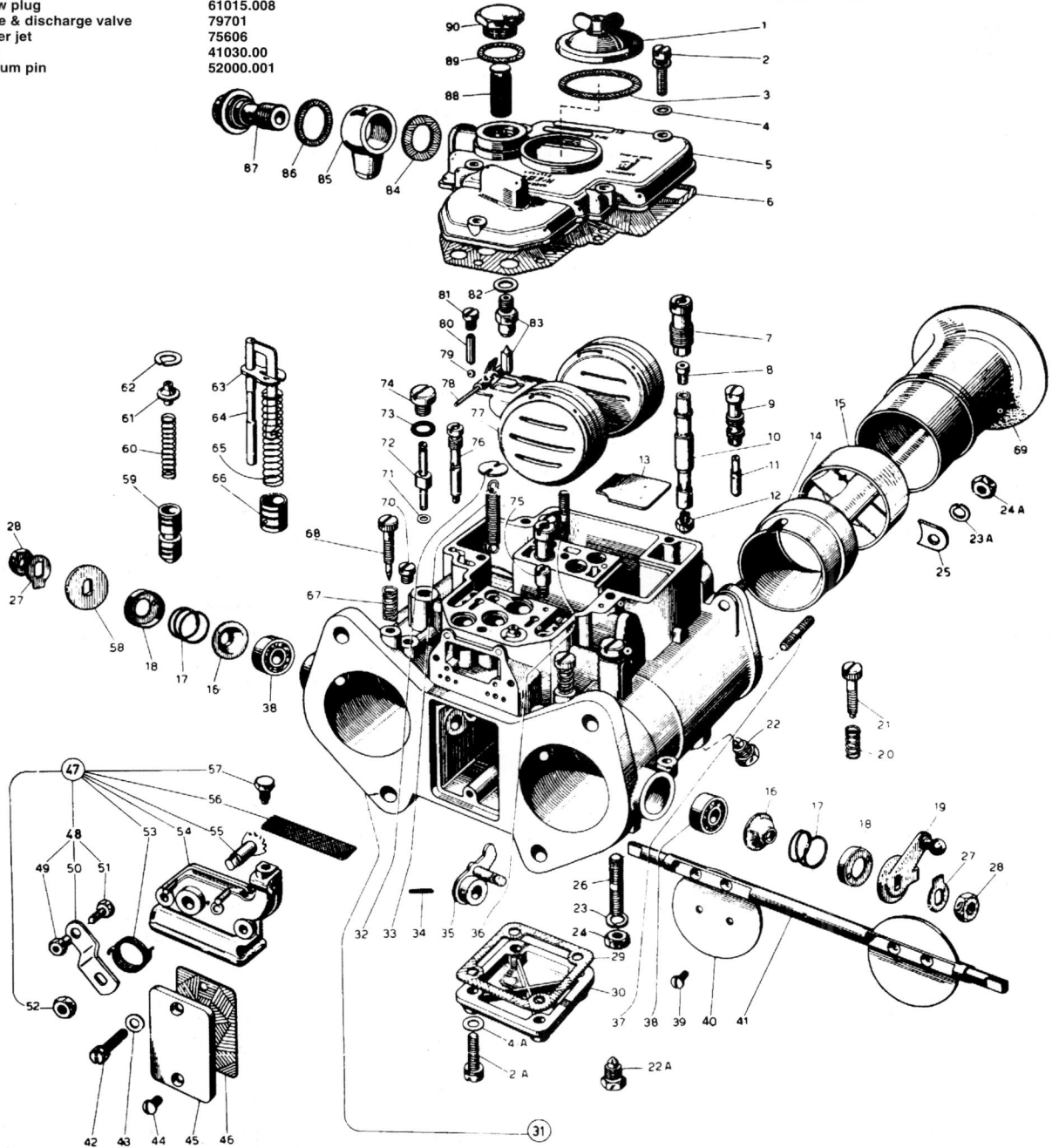

Comparison between 39mm full radius and standard type ram tubes.

Full radius ram tube with sleeve for carburettor choke.

Induction Technologies' (ITG) full radius ram in 39mm size just fits inside the filter case.

K&N filter with full radius rams and repositioned cover support pillars.

With the very large K&N filter, clearance at inner wing edge is tight.

a 1.53in (39mm) rams on my car and find that, for my engine specification, these work best. However, you should test different ram lengths during a dyno session to discover the optimum length for your car's engine.

Perhaps the best rams available currently are in aluminium, of full radius design and made by Induction Technology Ltd: they can sell you a pair direct or you can get them from most Weber dealers. Some fettling is required before fitting the full radius rams because, unlike Weber and K&N rams, they have a flat base and fit against the face of the carburettor rather than sliding inside the carburettor choke.

One way to ensure a good fit is to cut an old Weber or K&N ram down to size to provide the necessary sleeving inside the carburettor: the Induction Technology ram will then fit flush to the carburettor flange and the sleeve. An alternative to sleeving the carburettor when using the full radius ram, is to radius the Auxiliary venturi and the inside edge of the carburettor.

Do not be tempted to run your engine without an air filter as engine bore wear will be excessive. While running the engine without rams, will seriously degrade engine performance. If you do use full radius rams with a K&N filter, only the 9 x 5.25 x 3.25/4.5/5.5in (229 x 133 x 83/114/140mm) bodies will accommodate them. Even then, it's necessary to reposition the internal pillars that secure the filter case lid to the outermost edge of the case. It is thought that K&N can supply different stampings of the case or leave out these holes for you to drill yourself. The author drilled new holes in his case and blanked off the old ones with insulating tape.

A final point to note is that whatever ram pipe and filter case combination you use, make sure you have at least 1.18 inch (30mm) between the end of the ram and

in different lengths and, in my experience, the closer you can get to 2.55in (65mm), the better the engine will run and the larger the main venturi you can use. I use

The Weber DGV (DGAV) conversion for the A-series engined Sprite/Midget as retailed in the USA by Victoria British Ltd.

the filter case, especially on K&N cases, as less clearance than this will impede airflow.

WEBER 32/36 DGV (DGAV) CARBURETTOR CONVERSION

A popular conversion in the USA is to fit the Weber 32/36 DGV (DGAV) two barrel progressive downdraught carburettor. I have no experience of this conversion whatsoever but, from the photos I've seen, it does look as if bonnet clearance might be a problem, requiring a substantial bulge to clear the carburettor. Aside from possible height problems, however, this would appear to be a good conversion. The DGV (DGAV) is a very common and functional carburettor that's fitted to many Ford engines and may well work out a cheaper option to a DCOE conversion, given the secondhand availability of these carburettors. Victoria British Ltd in the USA is one company I know who retails this conversion kit. I would, of course, be interested to hear from any reader who has experience of this conversion.

SU HS4 CARBURETTOR CONVERSION

You can convert your car from the standard 1.25 inch SUs to 1.50 inch SU HS4 units. The carburettors and manifold are available from Janspeed Engineering who will also be able to advise on calibration for your particular engine.

If you're going to use K&N air filters on a Midget/Sprite with twin SU carbs, then this filter case is the one for you; this one's for the Midget.

Turbocharger adapted from Metro Turbo fitted to Simon Atherton's Midget.

FUEL-INJECTION CONVERSION

The author's not aware of anybody running a fuel-injected Midget/Sprite yet, but it seems likely that systems will be adapted to the cars in the near future by those seeking ultimate performance or needing to comply with particularly onerous emission standards.

It should be possible to fit Weber's Alpha aftermarket electronic engine management, fuel-injection and integrated ignition system to Midget/Sprite. I understand the system successfully reduces harmful emissions, reduces fuel consumption and improves overall driveability but doesn't make much difference to top end power. Unfortunately, the Alpha kits are expensive and must be fitted by Weber agents. Other makes of aftermarket fuel-injection systems are available and their manufacturers should be able to advise on suitability for any application.

FORCED INDUCTION

Turbochargers

It is possible to turbocharge the A-series engine by adapting parts from a Metro

Turbo: Simon Page's racing Turbo Frogeye is a good example.

Road cars with a similar turbo conversions exist, one being Simon Atherton's car which also uses Metro Turbo parts, including the camshaft (working with hi-ratio 1.5:1 rockers). Boost pressure on Simon's car is set at 7.5psi and can be adjusted from inside the car. Note that the standard Metro Turbo boost is a lowly 4 to 4.5psi, limited by a modulator valve until quite high up the rpm range. The engine has been modified to run a lower compression ratio: initially 7.44:1 but more recently down to 7.1:1. Simon achieved this by using more heavily-dished (16cc) pistons than standard and reworking the cylinder head which, although modified, still uses standard valve sizes.

This conversion not only requires the turbo, manifold and carburettor, but also the high pressure fuel pump, fuel regulator (which must be boost sensitive), return fuel line from the regulator back to the fuel tank, Metro Turbo head gasket, very hard (cold) sparkplugs NGK BP5ES(V), distributor modified to produce boost retard as well as vacuum advance and a boost gauge (optional). From this list you can see the easiest option is to get as many parts from a Metro Turbo as possible. Installation will be a case of following Metro Turbo workshop manual instructions, adapting as necessary for the Midget/Sprite. The bonnet will require some slight modification to allow clearance for the carburettor plenum chamber.

The point about turbocharging the Midget/Sprite is that no tailor-made kits seem to be available currently, so you'll need to adapt a turbo from a different model such as the Metro Turbo. If you fit a Metro Turbo kit, you might also consider fitting a Micro Dynamics upgrade kit which contains: a driver-controllable wastegate boost adjuster, ignition boost retard (which is fully adjustable), boost gauge to 15psi and a rev limiter. This company also makes a range of turbo ignition management packages, two of which contain a rev limiter. The turbo ignition management package can also be used on a supercharged engine.

Shorrocks & Judson superchargers

In the past, superchargers fitted on Midgets/Sprites were usually made by either Judson or Shorrocks (both of which

SU carburettor calibration for Shorrocks-supercharged cars				
Model	Carburettor	Needle	Jet	Dashpot spring
Midget/Sprite 948	H4 1.5SU	RG	100	Red
Midget/Sprite 1098	H4 1.5SU	BG	90	Blue
Midget/Sprite 1275	H4 1.5SU	RG(A)	100	Red

are vane-type units). The only Shorrocks-blown car the author knows of is David Clarkson's, which uses a C75B-type unit. David's car had the supercharger fitted by a previous owner so the precise origin of the unit is unknown: its likely to be a Speedwell or Allard kit both, unfortunately, no longer available. Some cars were fitted with a supercharger at the factory as original equipment!

Old Shorrocks units with part of the fitting kit do turn up from time to time and, should you be the finder of such unit and wish to fit it to your car, you may find the following points of help. The compression ratio of your engine - particularly if it is above 9.0:1 - may need to be lowered. You may also need to use 98 octane petrol and colder (harder) sparkplugs. I understand most Shorrocks kits were for

non-1275cc engines so fitting one to a 1275cc unit may require a special blower mounting plate. Boost should be set to between 5 and 7psi. Generally speaking, you are likely to a find 1.5 inch SU carburettor and carburettor pipe fitted to the supercharger, though other sizes were used. Check fan to supercharger drivebelt clearance, since the original kit used an aluminium spacer to bring the fan clear of the belts. Disconnect the vacuum advance pipe on the distributor, ensuring the inlet end is plugged at source in order to prevent an air leak, which would cause lean running. Otherwise, fitting should be straightforward, assuming you have all the parts, since the Shorrocks unit was designed as a bolt-on kit which didn't require other engine modifications. The carburettor calibration (assuming an otherwise standard engine) for the Shorrocks unit is given in the accompanying table.

Once the Shorrocks supercharger is fitted and running satisfactorily, the only maintenance required is to remove the lubricator pin every 1000 miles and wipe it clean with a soft rag. UNDER NO CIRCUMSTANCES SHOULD ABRASIVE BE USED FOR CLEANING. Removal procedure is as follows: unscrew the pipe connection at the supercharger (whereupon the pin will be partially pushed out by the spring positioned underneath it); insert a small screwdriver into the slotted end of the pin and draw it out. If the spring should come out with the pin, remember that this must go in first when reassembling.

The recommended maximum sustained engine speed the supercharger

David Clarkson's supercharged 'Sprout,' a green Frogeye with Shorrocks supercharger.

Judson supercharger complete with inlet manifold and carburettor *(Courtesy Dean Hedin)*.

Mini Mania supercharger kit fitted to a Midget/Sprite engine. This kit will be available through Mini Spares in the UK. *(Courtesy Mini Mania).*

can be run at is 6000rpm (up to 7250rpm is okay for a short duration). Since the supercharger is direct drive by 4.5in (114mm) diameter pulleys, the supercharger rpm is the same as engine rpm.

All the Judson-equipped cars I know of are in the United States of America. One such is owned by Dean Hedin who helped me research this section. The Judson is straightforward to strip down, starting with the manifold, idler pulley and bracket and front endplate (using a puller, if necessary). The rotor assembly and back endplate can now be withdrawn from the casing and the rear end separated from the rotor. Further disassembly will be necessary to check the bearings for wear. The rotor housing must be inspected for wear and, if badly scored, will have to be honed out in much the same way as an engine cylinder bore is honed. If the endplates are badly scored they will need to be reground. Dean advises me that a plastic material known as Delrin is the best material for new vanes. The clearance between the endplates and vanes is 0.015-0.020in.

Judson superchargers are generally equipped with a single barrel Holley carburettor for which rebuild kits are still available. Dean has modified his unit to take a SES low pressure fuel injector.

The Judson unit's lubrication is by a total loss oil system. The method for total loss oil lubrication is to use a device known as a top cylinder lubricator which is made by the Marvel Mystery Oil Company. The best oil to use is a good synthetic.

The Judson charger should produce 6psi boost on a standard pulley which, like the Shorrock unit, is the same size as the crankshaft pulley and so produces identical rpm.

Mini Mania supercharger

Mini Mania recently released a supercharger kit for the Midget/Sprite which had undergone limited testing at the time of writing. The supercharger has a twin screw design, positive displacement compressor which is said to be much more efficient than other designs of supercharger.

The kit consists of a supercharger (appropriate to engine size), special inlet manifold, crankshaft pulley/ damper, water pump and alternator pulleys, drive belt, SU HIF6 carburettor (1.5 inch), K&N air filter and all necessary brackets. Because the kit is so new it is not possible for me to quote power figures for an otherwise standard 1275cc Midget/Sprite because they simply don't exist yet. However, it seems reasonable to expect that, when used with the custom-ground Elgin cam, power could be on a par with a very hot

Mini Mania supercharger in kit form for the Sprite/Midget. *(Courtesy Mini Mania).*

race engine. It's possible the kit may benefit from a slightly larger carburettor and that engine water and oil cooling would need attention in order to maximise horsepower. Finally, the author would strongly recommend using a non-standard gearbox and drivetrain as the standard items are unlikely to be strong enough to cope with the torque generated by a supercharged engine.

THROTTLE PEDAL

All non-1500 cars have the hinge-on-floor-type ('Organ') accelerator pedal. These

Pendent-type accelerator pedal from 1500 Midget fitted in place of earlier organ-style pedal.

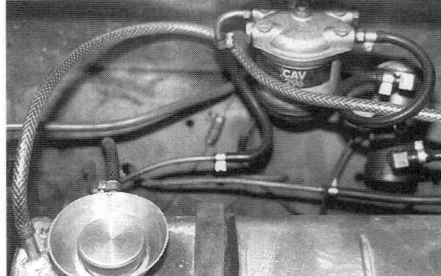

Boot-mounted competition fuel tank with Lucas fuel filter and an interrupter fuel pump. A high pressure pump with regulator in engine bay completes the fuel delivery system of Simon Page's turbo Frogeye.

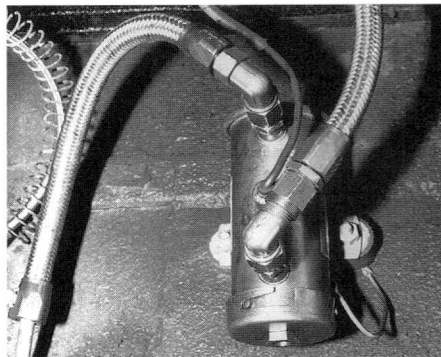

FSE Bendix-type fuel pump mounted on rubber mounts on author's car; Goodridge hose and fittings are put to good use.

pedals can jam (the throttle stays fully open) when pressed hard. Part of the problem seems to be in how much pedal travel is required to fully open the throttle; the longer the travel, the more likely the pedal is to jam.

The solution is fit the throttle pedal from the 1500 model. This model's pendent-type accelerator pedal is a straightforward swop. The only slight problem I found was that the cable adjustment on the throttle linkage needed to be backed-off as the new pedal was otherwise partially opening the throttle even though the pedal was 'at rest'; I'm not sure why this was, but it was easily cured. The two bolt holes in the floor left over from the hinged pedal fitting can be plugged with rubber grommets or welded.

FUEL PUMPS

If you have modified your car's engine to produce more power, you'll have discovered that more fuel is required to produce that power. The author's own,

once parsimonious, Sprite now delivers an average 25mpg; but then it does have considerably more performance than a standard car.

If your car's modified engine is not to suffer from fuel starvation (symptoms may show during dyno testing), you'll need to uprate the fuel delivery system capacity

until it can easily cope with the maximum fuel flow required.

The standard electric pump can only cope with moderate increases in fuel demand, and that's assuming it's in good working order. Depending on the state of your engine's tune, the easiest, and cheapest, way to ensure good fuel supply may be to fit a brand new standard pump. In some cases a similar pump from a different model with a larger engine may have sufficient capacity to cope with the extra thirst of your modified engine. Another solution is to fit two standard pumps to run in series.

For a really serious horsepower increase the answer is to fit a competition fuel pump. On 1500cc-engined cars the mechanical pump can be discarded and the opening for the pump fitting in the block blocked off or used as an additional engine breather - either way, this will require fabrication of a small plate. On A-series-engined cars the high capacity competition pump will simply replace the existing electric pump.

Shopping for a competition pump can be more than a little bewildering. One of the things to watch with a pump is the pressure it generates: just because the pump can produce 7psi (48.3kpa) - which doesn't sound much - it doesn't mean your car's carburettor can handle it. My suggestion is to 'phone a pump supplier and ask for advice. One company I can recommend is Facet (UK) Concessionaries (FSE), the main importer of the range of electric fuel pumps that just about every-one will try to sell you.

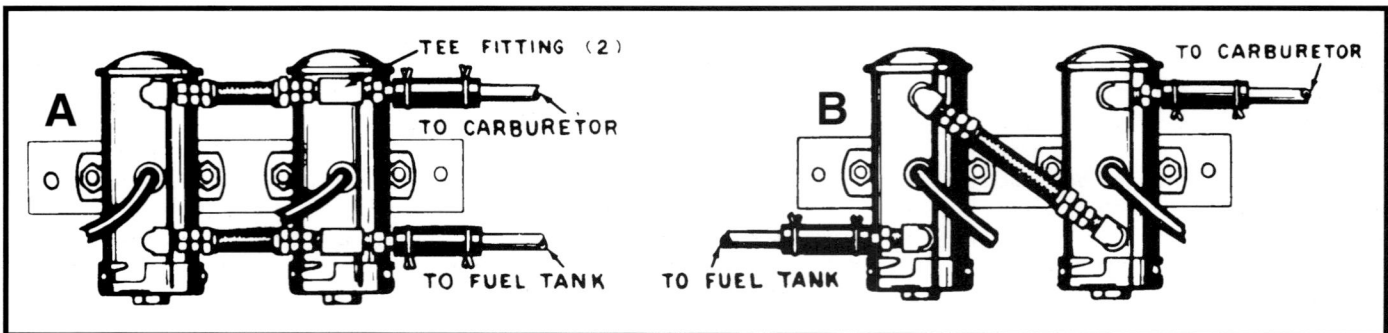

Fuel pumps can be plumbed in parallel (A) for twice the fuel flow at the same pressure, or in series (B) for twice the pressure and normal flow. (Courtesy Facet Enterprises Inc).

FUEL REGULATORS

Establishing which is the best pump for your engine isn't the end of your problem. Because of the high pressure output that goes with the high flow pump you will most likely need a fuel regulator as well, if you haven't already got one fitted. The regulator will have a pressure adjusting facility which should be set to suit the fuelling requirements of your carburettor/s. Your retailer, or Facet, can advise on the best setting for your particular application.

Facet fuel pump situated adjacent to the fuel regulator in the back of Dave Grove's racing Frogeye.

Filter King fuel regulator and filter combined.

Facet supplies two types of regulator: one with push-fitting connections for normal petrol hose, or a more recent type that has threaded fittings. The threaded fitting type is to be preferred as it will allow the use of braided metal hose for the fuel line. Braided metal hose may sound like a luxury for the fuel line but, believe me, it's not.

If your tuning route was by way of either supercharging or turbocharging,

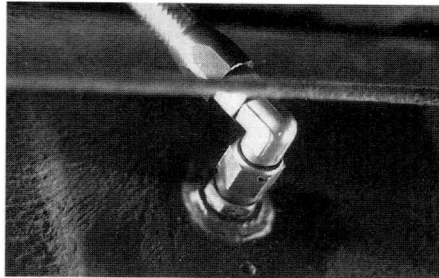
Goodridge hose and fittings at fuel tank union on author's car: this was the only way he could think of to run a large bore fuel line right up to the tank.

you'll have a whole host of other considerations concerning fuel pressure and induction boost. Not least of these will be whether you need to have a special seal kit for the carburettor. There is a wide choice of variables to consider and, again, I suggest you seek expert advice, having full details of your modifications to hand.

With your new high capacity pump and regulator (which is also likely to incorporate a filter), you're almost there. A final small, but vital, consideration is the bore of the petrol pipes, both rigid and flexible. There's not much point spending a lot of money on an efficient pump and regulator only to restrict fuel flow with skinny pipes. Choose a realistic pipe bore according to your engine's needs. A lot of pumps have special unions and I assume your retailer sold you the relevant ones to go with the pump. However, it may be worth considering a braided metal hose and union kit. Goodridge can supply your needs if you give them details of fittings and length of hose required.

FITTING

So finally, it's time to fit the pump, fuel lines and regulator to the car. Ideally, the pump should be fitted as near the fuel tank as possible and this is not difficult on the Midget/Sprite as siting can be quite near the original electric pump. On 1500cc cars, try to fit the pump as near the fuel tank as possible. It may seem like a good

idea to fit the fuel pump under the bonnet but, in practice, this does not work as well: the reason for this is that the engine bay air temperatures are on the warm side and can cause fuel vaporisation problems. Where you finally decide to mount the pump will be up to you, but make sure you follow the guidance contained with the pump fitting instructions. Pumps can be noisy and some models are available

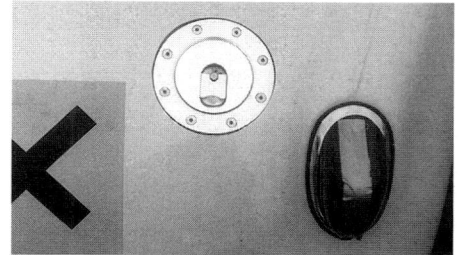
Flush fit fuel filler cap on Dave Grove's racing Frogeye looks neat.

with noise reducing rubber mountings.

Once the pump, pipes and regulator are installed the only remaining job is to wire the pump. It may be a simple job of using the existing pump connections but, if you are wiring the pump to a separate switch and not through the ignition, be sure to mark that switch(es) suitably. FSE can also sell you a safety shut-off switch: a single pole, double throw pressure switch which is connected into the pump's electrical supply. Activation is by engine oil pressure, so stalling the engine shuts off the pump. This is worth considering if you are going to do a lot of motorsport with your car or simply as an additional safety feature. Another piece of electronics which may be of interest is an engine rpm-switched relay. This is a 'black box' from Microdynamics that can be used to switch on an auxiliary fuel pump (for the racer) or a nitrous oxide system.

With the new fuel system parts fitted run the engine (outdoors) and check for leaks. Then, and only then, test the car either on a rolling road dynamometer or race circuit to ensure fuel starvation has been eliminated.

Chapter 3
Ignition System

MODIFICATION OPTIONS

Although even a modified ignition system cannot deliver much in the way of new horsepower, a poorly maintained or inadequate set-up will always lose horsepower. Perhaps more importantly, an efficient ignition system *will* provide good starting combined with reliability and minimum maintenance.

When in good condition the standard ignition system is more than adequate for the standard engine. However, an engine tuned or modified for greater street performance, or to competition standard, is more demanding in its ignition requirements. The larger volumes of fuel/air mixture (often at higher compression) that are burnt in tuned engines will require a higher voltage at the sparkplugs to ensure complete combustion, and higher engine speed (rpm) requires more sparks per minute.

The total voltage that can be delivered to the sparkplug is the main factor governing whether or not the plugs fire successfully. Voltage shortfalls at the plug can be attributed to coil output, losses due to resistance in the HT leads and, finally, the sparkplugs themselves can also limit spark quality and strength if their initial resistance is too high.

The accompanying chart shows how to match ignition system specification to intended use.

DISTRIBUTORS

In the USA a popular conversion from the standard ignition system is the Mallory dual point distributor. This is a mechanical advance distributor that has the advance curve pre-set at the factory. The dual stabilised points eliminate points bounce and erratic timing that are a common feature of the standard distributor, especially a worn item used at high revs. However, the key value of the dual points is the increase in coil saturation time they create which boosts the output voltage (spark) no matter what coil you are using.

In addition to the dual point distributor Mallory also produces a breakerless electronic distributor and ignition system. There are two specific models to choose from. The first is the Mallory 45-series Unilite model in which triggering is by a self-contained photo-optic infrared light emitting diode system. This distributor can be used with standard or uprated coils.

Intended usage	Check sparkplug heat range	High-performance ignition (HT) leads	Electronic ignition system	High-output sports coil	Rev limiter
Standard or mild road	No/yes	No	Desirable	No	No
Fast road	Yes	Yes	Yes (contactless best)	Yes	Desirable
Competition	Yes	Yes	Yes (contactless best)	Yes	Yes

The advance is mechanical. The second is the Mallory 47-series Unilite with vacuum advance. In all other respects it is similar to the 45-series unit. On both 45 and 47 models, the advance curves are adjustable and it is this factor which makes them ideal for the out-and-out racer. They will be suitable for some high-performance road cars too, though, for road use, the advance curve may, in some instances, be less than ideal.

DISTRIBUTORLESS IGNITION.

Huffaker Racing has a kit (retailed through Mini Mania, USA) that converts A-series-engined cars from distributor ignition to a flywheel-triggered ignition unit.

Here's the sensor and rotor of a Huffaker Racing distributorless ignition system.
(Courtesy Mini Mania).

ELECTRONIC IGNITION SYSTEMS

The advantage of so-called 'electronic' ignition systems is that they do away with most of the vagaries associated with mechanical (points and advance mechanism) spark timing. However, electronic systems come with many different features. The author has used the Microdynamics Formula One electronic ignition system for some years now: it was originally chosen because it was a contact breaker points-triggered system and could, therefore, be fitted to the car without altering the distributor advance

Micro Dynamics Formula One contact breaker points-triggered electronic ignition system with built-in rev limiter.

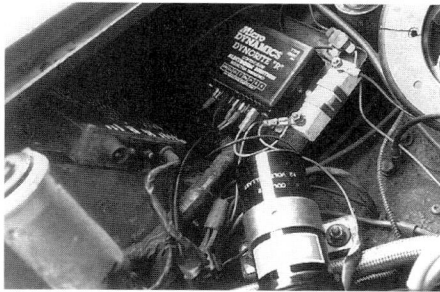

The Formula One system installed on the inner wing of the author's Sprite. The Dynorite contactless ignition module is on the pedal box cover.

Microdynamics fully electronic, contact-breakerless distributor.

mechanism. Please note that Autocar Equipment Ltd., which retails the Lumenition range of ignition products, has bought out most of the Microdynamics range of ignition products (except distributors) and can assist with any ignition problem.

If you have a highly-tuned engine in your car, you may well have moved on to a different distributor such as an Aldon or Microdynamics unit which may have had its advance curve adjusted to suit your

Ignition module for Crane Cams contact-breakerless ignition convertion.

engine's state of tune and ignition system. Since original fitment, the author has progressively upgraded the Microdynamics system on his car into a fully-electronic package; this involved having the distributor rebuilt and another ignition package fitted to drive the distributor electronics in conjunction with the existing system. During the rebuild of the distributor, the opportunity was taken to have its ignition advance and retard system re-curved to take advantage of previous rolling road information on the ignition advance demands of the car's tuned engine . When you read the rolling road chapter you'll find further comment on optimising ignition curves.

Usually the coil can remain in its original location. I have mounted ignition system 'black boxes' both on the pedal box cover and inner wing without any problems, but, bear in mind, that the main considerations for siting ignition electronic units are to keep wiring runs short and the units away from excessive heat, dampness or vibration.

Ignition system modifications to my own car started off with a Microdynamics coil and distributor, then a rev limited points-triggered system and, finally, a contactless system. One of the beauties of the Microdyamics range (and some other manufacturer's ranges) is that there is scope to expand and upgrade the system as the need develops.

Another electronic ignition system - though this time a conversion kit for the standard Lucas distributor - is available from Crane Cams. The Crane kits suitable for the Midget/Sprite are: XR200, XR2000

Optical pick-up and chopper wheel (1 surplus) and fittings for the Crane ignition system.

and XR3000. Note that descriptions of these kits are likely to suggest maximum rpm is in the range of 6000-6500 but this only applies to V8 engines and they will run any revs you can wind an A-series engine up to. Triggering is by infrared light beam. Fitting the kit is quite involved but the instructions are comprehensive. Once the system is set-up, it requires no further adjustments. Crane's instructions also contain a trouble-shooting section which is worth keeping for future reference.

In the UK, a popular breakerless ignition conversion is the Aldon Ignitor which, unlike most conversions, does not require a separate 'black box' but fits entirely within the standard distributor body. I have had good reports of this system from both Peter May Engineering and racer Simon Page.

IGNITION (HT) LEADS

HT leads have some resistance to electrical current. Most Midget/Sprite leads are manufactured from carbon impregnated string, which has the highest resistance value and which, after about 20,000 miles (30,000km) no longer functions very efficiently. For performance use, the best of the available alternatives is the wire-wound, inductively-suppressed HT lead and, as direct replacements, such leads deliver the highest energy to the plug.

Amongst the better grades of ignition HT leads I can recommend are those supplied by Micro Dynamics Ltd., NGK Ltd., Mini Spares and Crane Cams.

SPARKPLUGS

Every sparkplug has electrical resistance due mostly to the gap between the electrodes. However, the high tension (voltage) current passes through the central electrode which, in standard plugs, is made of nickel alloy which has significant electrical resistance. An electrode with less resistance will produce a stronger spark, so both NGK and Bosch manufacture plugs with copper or platinum central electrodes which are excellent electrical conductors with minimal resistance.

Very important to engine and ignition performance is the grade of spark plug you use. For the record, the author's preference is for the NGK BPES or BPEV-series of plugs, though he has found that it is not possible to get much more than 9000 miles (14,500km) from a set and, apparently, this is only to be expected when plugs are used in a high performance engine with a good ignition system. High performance engines like colder (hard) rated plugs.

To find the optimum heat range plugs for your car, start with coldest plugs you think might be suitable (say, two grades colder than standard) and then work towards warmer ones; this way no harm will come to the engine. Note that Champion and NGK use numbers that progress in the opposite direction to heat range, so be sure the tuning shop understands your requirements. Plugs which are too hot for the engine will create pre-ignition (knocking/pinking) after or during hard use and may cause the engine to run-on after it is switched off. Prolonged use of plugs which

NGK BP8ES and BP8EV plugs: note protruded tip which is indicated in the plug marking by "P." (Courtesy NGK).

are too hot can damage the Midget/Sprite engine.

SPORTS COILS

By making the best of plug and lead choices the ignition system is now capable of delivering the full potential of the standard ignition coil. However, to produce a higher voltage, a high-output sports coil is required. High-output coils have a more secondary windings than standard and produce a greater voltage as the contact breakers open.

Several manufacturers produce sports coils suitable to use as direct replacements for the standard item. The drawback with a sports coil is that it takes more current and can increase contact breaker points arcing, leading to higher erosion and the need for more frequent resetting and renewal of contact points.

The author has used both a Lucas sports coil and two different Microdynamics coils on his car's engine, all with good results; although the Microdynamics coils seemed to have the edge over the Lucas on output. Other high-output coils are available from Mallory and Crane.

Lucas sports coil.

The oddly-shaped Crane high-output ignition coil is fine for any power-tuned Midget/Sprite.

REV LIMITER

This device cuts ignition pulses at a pre-set engine speed to prevent the engine over-revving and consequent mechanical damage. Microdynamics produce an engine cut-out that is available either separately or incorporated into an electronic ignition system. It's called 'Smooth Cut' because it works electronically rather than mechanically and will not cut the ignition completely when it comes into operation.

A new and inexpensive rev limiter has recently been produced by Armteca. The rev limiter is a small unit which will clip onto any round body-type of coil and is easily wired. Once fitted, it can be set to your required rev limit. Although the author has no experience of this product, it has received favourable reviews in at

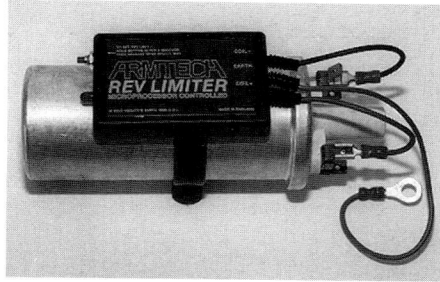

Armteca rev limiter mounted on an ignition coil. (Courtesy Armteca).

least one leading car magazine.

RPM TELLTALE

If you have one of the fine tachometers from Stack then you will have a telltale feature built into the unit which can be utilised by the use of two separate and remotely positioned function switches. If

A selection of black boxes: speed shift, rev limiter and telltale - all from Armteca. (Courtesy Armteca).

you don't have a Stack tachometer, or even another make with a telltale function, you can purchase a 'black box' telltale from Armteca. As with other products from the Armteca range, it is supplied with good fitting instructions and is easy to install.

Chapter 4
Cooling System

COOLING SYSTEM GENERAL

Of the heat energy produced by an internal combustion engine, about a third is dissipated by the cooling system. Without this cooling the engine would overheat, and its internals would seize. An engine tuned to produce more power will also produce more heat and, whilst the Midget/Sprite cooling system can handle moderate increases in engine power before becoming overstretched, you'll need to uprate the system for serious high performance use.

For maximum engine power the best coolant temperature is about 158 degrees F (70 degrees C), which is quite a bit lower than normal but, unless you're intending to us the car for competition, it's not necessary to achieve this low a running temperature.

There are many ways to modify the cooling system, and the options you select will depend on how you intend to use your car. A car that is both raced and used as a road car may have to have a compromise system because the best set-up for racing may well mean road temperatures are on the cool side, though this could be remedied with a radiator blind. In winter, with a very efficient cooling system, it's possible to end up with an engine running too cool; the remedy, again, is to partially blank-off the radiator.

Finally, if the radiator or hoses are leaky and suspect, you'll never succeed in keeping water temperatures down, so start with a good, well maintained system and you'll achieve good results. Remember that failure to achieve proper cooling for a high performance engine could result in permanent damage.

RADIATOR(S)

Lowering the temperature of the engine's coolant is best achieved by increasing airflow through the radiator (or a bigger radiator for the same effect), improving the speed and volume of coolant circulation through the engine or, to a lesser extent, increasing coolant capacity. Auxiliary radiators, which will increase coolant capacity and expose the coolant to more airflow, are available from manufacturers such as Serck, or you could use an old heater element to increase the radiator-area of your system.

Simon Page's turbocharged Frogeye Sprite used an Austin Allegro radiator originally but, as Simon gradually coaxed a lot more power from the engine, it became apparent that the engine water temperature was on the high side. Since Simon already had an auxiliary radiator fitted, the solution appeared to be a larger capacity radiator. The author approached Serck's competition division on Simon's behalf and, after expert advice from Mel Gosney, Senior Manager, a specification was agreed. Serck build competition radiators for everyone who is anyone in motorsport, from club racers through to the Works rally stars, as well as for the roadgoing Jaguar XJ220.

When choosing a non-standard radiator, pick the unit with the largest possible surface area. Peter May Engineering used a bigger radiator in its old racing Midget to good effect. If it is not possible to use a radiator with a surface area larger than standard because of space restrictions or a wish to preserve originality, then

Serck aluminium racing rad fitted to Simon Page's racing turbo Frogeye.

Serck aluminium racing rad adjacent to the turbo intercooler in Simon Page's turbo Frogeye.

Serck aluminium racing Sprite/Midget radiator fitted to the author's Sprite. Note Kenlowe electric fan.

increase coolant capacity by using a thicker core.

The next choice is between conventional copper and brass construction or aluminium. The typical weight saving when switching to aluminium construction is 2lbs (0.907kg). Serck aluminium cores also have a higher density of water tubes than a conventional core which produces an extra 3 tubes per 6 inches (15cm). On any given Serck alloy radiator the gain in the number of water tubes alone produces a unit which is substantially more efficient than a conventional unit. Engine testing on

Speedwell Engineering aluminium radiator. *(Courtesy Tom Colby, Speedwell Engineering).*

A Speedwell radiator fitted to Tim Walker's Sprite. *(Courtesy Tom Colby, Speedwell Engineering).*

a chassis dyno with the Serck aluminium radiator fitted to Simon's car was a great success: the new unit proving vastly more effective than the old radiator and auxiliary radiator combination.

When the author switched to a Serck competition radiator to replace his car's standard crossflow radiator he was very impressed by the new unit. Town driving (with existing Kenlowe fan), produced a running temperature 41 degrees F (5 degrees C) cooler than with the standard radiator. However, it was during fast driving conditions that the more impressive gains were made. At an approximate speed of 60mph (3500rpm in 4th gear) average temperatures still lower by around 41 degrees F to give about 185 degrees F (85 degrees C). Once speed increased to 75mph and on to 95mph (4500-5500rpm in 4th gear) the temperature fell sharply to just under 176 degrees F (80 degrees C) and kept falling until I slowed down. I have no doubt that sustained high speed driving would have produced even lower tem-

peratures. This is, of course, the reverse of my experience with a standard radiator. A.P.T's David Anton reckons on optimum horsepower occurring at around 158 degrees F (70 degrees C) so any radiator which can achieve this is worth some extra bhp. A Serck competition radiator can save you weight and increase cooling capacity for the same external dimensions as a standard unit.

Speedwell Engineering in the USA also retails an aluminium radiator for the Midget/Sprite with 15 water tubes per 6 inches (152mm), each 1.25 inches (31.75mm) deep. Aside from dropping water temperature more than a standard radiator, it is also lighter at 4.5lbs (2.04kg). I would expect this radiator to perform in a very similar manner to the Serck competition radiator.

In Australia, Gillspeed retails a special super-efficient-cored radiator.

RADIATOR AND EXPANSION TANK CAPS

The Midget/Sprite cooling system operating pressure is determined by the radiator cap, and different models have cooling systems with different operating pressures. The range is from 4psi (27.58kpa) to 15psi (103.5kpa), although the most common rates are 7 and 13psi (47.26 and 89.63kpa). By pressurising the system the coolant's boiling point is raised. Each pound per square inch (6.9kpa) of pressure, will raise the boiling point by 3 degrees F. Not only does this allow the running temperature to rise until well above boiling point, it also allows the use of a smaller radiator.

The author uses a 13psi (89.63kpa) radiator cap (the standard rating for the model) on his Sprite cooling system. If you have an earlier model, for which a lower rated cap is specified, you can use a higher rated cap to provide a safety margin. (Once coolant has been expelled from the system's overflow in heavy traffic or sitting on the start line at a race, it cannot be replaced until the system has had time to cool down. Once on the move again, a system with reduced capacity is even more likely to run hot, which will lead to further coolant loss, etc.).

If you do use a higher rated radiator cap than standard it can create problems for the system by blowing out weak core plugs in the cylinder block or causing leaks at hose joints. The answer is to make sure the cooling system in perfect order before fitting a higher rated radiator cap.

THERMOSTATS AND BLANKING SLEEVES

There is a thermostat in the cooling system to allow the engine to reach operating temperature quickly by cutting off water circulation to the radiator until the coolant reaches a predetermined temperature.

It's possible to get a thermostat that will open at between 149 degrees F (65 degrees C) as opposed to the more usual 158 degrees F (70 degrees C), or even as high as 176 degrees F (80 degrees C). A change to a thermostat that opens at a lower temperature will tend to reduce the normal running temperature of an engine, providing the radiator's cooling capacity has been increased as described earlier.

Alternatively, the thermostat can be removed and a blanking sleeve substituted. At the same time, remove the bypass hose and plug the stubs on the head and water pump. Unless you have the cylinder head off the engine, the author suggests fitting blanking plugs to two bypass hoses and fitting one hose to the water pump, the other to the block. Blanking sleeves and plugs can be ob-

tained from Mini Spares and other A-series engine specialists.

The point of the so-called 'blanking sleeve' is not to block the flow of water but, instead, divert it to create better distribution of coolant around the whole engine.

HEATER AND HEATER TAP

An option when modifying the cooling system is whether or not to retain the heater element. I consider retention unnecessary when the cooling system has been modified by fitting a blanking sleeve in place of the thermostat; even without the element, the heater blower can extract enough warm air from the engine bay for demisting the windscreen.

Racers may wish to keep the heater element but, for reasons outlined in competition preparation chapter, it will be necessary to relocate the unit.

If you do remove the heater element, retain the heater tap but reverse its position so that it will be easy to run a hose from the tap to the union at the front of the engine. NEVER turn off the heater tap as this action will cause overheating of number 4 cylinder.

Direct hose fitting in lieu of heater tap, this hose bypasses the heater matrix (which has been removed from heater box).

WATER PUMPS

With the A-series engine there are quite a few alternative water pumps to choose from as most of the later Metro pumps and most Mini pumps will fit the Midget/Sprite,

Standard 1275 water pump (right) alongside Mini Spares Metro pump; note deeper impellor.

Standard (left) and Mini Spares Metro pump. Note removal of bypass stub on the Mini Spares unit.

although it's not necessary to get a high capacity water pump just because you have a high power output engine. The requirement is for the smallest possible capacity pump that will get the job done and sap the least amount of power from the engine.

It is suggested that you retain the standard pump and only fit an alternative if all other methods of increasing cooling capacity fail to give the desired result. Conversely, if the engine is over-cooled, you could fit a pump of a capacity smaller than that of the standard unit.

If you have modified your car's cooling system by removing the bypass hose and fitting a blanking sleeve, you'll have already blocked off the water pump bypass outlet. If, however, you purchased a late Mini or Metro pump, you'll find that the bypass outlet is cast blank. The author discovered that this part of the casting is solid (on his Mini Spares-supplied pump, at least), so he cut it off and ground it flush

A-series engine blanking sleeve.

with the rest of the casting because he thought it looked neater.

PULLEYS

The standard size water pump pulley is 3.9 inches (99mm). On highly-tuned road or racing cars using the A-series engine it is possible to fit a reduced speed water pump pulley in the same way as a reduced speed dynamo pulley can be fitted. The main reason for fitting a larger pulley is to increase efficiency of the water pump at high rpm. The standard size pulley can run too fast and cause cavitation and quite severe frothing in the coolant at higher engine speeds, reducing cooling system efficiency by quite a margin. Using a larger pulley drastically improves this situation; typically, a temperature drop of around 41 degrees F (5 degrees C) can be expected (using the latest large impeller pump on a 1275cc engine).

Another benefit of a larger pulley is a reduction in the power required to drive the pump. The cooling system acts like a small water-brake: a more favourable ratio means more power available to drive the car. The difference is measurable on a dynamometer. The reduction in pump speed at low rpm, after fitting a larger pulley, is very unlikely to cause overheating problems. Mini Spares retails large water pump pulleys in two different sizes: 4.75in (121mm) and 4.5in (114mm) diameter. American and Canadian cars fitted with the exhaust emission control system air pump will not be able to use the Mini Spares pulley if the air pump is retained. I have not found a reduced speed pulley for the 1500 engine.

Standard (left) and large diameter (larger of two options) water pump pulleys.

ENGINE-DRIVEN COOLING FANS

If you want to reduce coolant temperature by increasing airflow through the radiator during traffic/slow speed driving, the best way to do so is to fit a more effective cooling fan.

An engine-driven fan nearly always produces the highest airflow figures, and the range of engine-driven fans for the Midget/Sprite go from two- to sixteen-bladed. An electric fan will never be as effective as a multi-bladed, engine-driven fan at high engine rpm, but the electric fan will push or draw more air through the radiator when the engine is ticking over in traffic or running at low rpm.

It is possible to keep the engine fan and have a Kenlowe electric fan for auxiliary cooling, but more about electric fans later.

ELECTRIC FANS

Having reduced coolant temperature by one, or more, of the previously described methods, the engine-driven fan is generally made redundant because the engine is cool enough, except in slow traffic. Fitting an electric fan, in place of the mechanical one, will correct the latter problem because it only switches on when required (via a thermostat or driver-controlled switching). An added bonus of using an electric fan is that removal of the engine-driven fan can release an extra couple of bhp.

Kenlowe is a well-known manufacturer of electric fans and I have used one of the fans on my car for some years without experiencing any operational problems. My fitting is unusual since it is a 12in (305mm) diameter fan whereas the kit Kenlowe supplies for the Midget/Sprite is normally 10in (254mm).

Standard Kenlowe fans for the Midget/Sprite come with comprehensive fitting instructions, but do not contain any specific points relating to the Midget/Sprite. Kenlowe offers two types of fan mounting bracket: V-mounting that attaches the fan direct to the radiator

Kenlowe electric fan. Note the adjustable bars which support it. *(Courtesy Kenlowe).*

Electric fan fitted to remains of bottom radiator cowl.

matrix (Unifan type) and adjustable bars (thermostatic). Whatever the application, the author strongly advises that the adjustable bars mounting system is used, even though it can take a little longer to fit. When you order your fan, request that two extra plastic C-clamps and fittings are supplied as these extra clamps ensure a more rigid fitting than standard.

Fitting for cars with vertical flow radiators is much simpler than for cars with the later crossflow radiator (late 1275- and 1500-engined cars). On vertical flow radiator cars, the bolt holes for the radiator shroud can be used to fit the C-brackets which will hold the adjustable brackets in place: bolts longer than those in the kit will be required to do this. This makes for a very neat installation but leaves only minimal clearance for an oil cooler (where fitted).

Goodridge metal braided hose can be used for water lines, too.

Crossflow radiator cars may be able to use the radiator shroud bolt holes as mountings, but it may be just as easy to make new holes. An alternative to using the 10in (254mm) fan on crossflow radiator cars, is to fit the larger 12in (305mm) fan and use fabricated brackets, eliminating the radiator shroud altogether. This can be done by removing the radiator and shroud, then cutting the shroud back to leave just the bottom section and cutting a slot in the bottom to allow clearance for the fan; alternatively, you can fabricate your own brackets. The top of the radiator will need to be supported by fabricated brackets (simple to make from sheet steel/aluminium) anyway. This done, you can refit the radiator. Not only have you now made room for the larger sized fan, but also increased airflow to the engine bay.

Just in case it isn't obvious, the fan must sit in front of the radiator. Once the electric fan is fitted securely, the old engine-driven fan can be removed. The pulley that the fan was attached to MUST, however, be retained because it drives the water pump. Note that you MUST use considerably shorter bolts than the originals when refitting the fan-less pulley, regardless of whether or not you retain the spacer. Failure to do so will cause the bolts to jam against the water pump body.

The Kenlowe fan fitting instructions contain a wiring diagram which is straightforward to follow. You have the choice of wiring the fan to the thermostatic switch, to an illuminated switch, or both. When you wire the fan, it is recommended that the electrical connections are soldered as well as crimped to ensure they remain secure. The siting of the fan switch and the thermostatic switch will be a matter of personal preference.

The Kenlowe fan kit comes with a thermostatic switch for automatic operation; however, the author's preference is for manual control via a dashboard-mounted switch which is illuminated when in the 'On' position. With a manual switch you can start the fan running the moment you enter traffic rather than wait until the engine gets hot enough to operate the thermostatic switch.

Whichever switching method you employ, be sure the fan is pushing air through the radiator. If the electrical connections are reversed the fan will try to suck air through the radiator against the airflow of the car and water temperature will soar!

Finally, if you run wire through any part of the car's bodywork, such as the engine bulkhead, use a rubber grommet to protect the wire. Failure to do this will eventually result in a short circuit and possibly even a wiring fire.

Chapter 5
Exhaust System

TUBULAR MANIFOLDS

For the Midget/Sprite the most obvious modification to the exhaust system is to discard the cast iron manifold and replace it with a long centre branch tubular steel item. All aftermarket performance manifolds are made from tube steel and, for this reason, they're usually known as tubular manifolds.

Fitting a tubular manifold may create problems in that it can produce more radiant heat than a standard cast iron manifold and, because of this, engine bay temperatures will rise. The A-series engine intake and exhaust manifolds are on the same side of the engine so, in heavy traffic, the fuel in the carburettors has a tendency to vaporise if underbonnet temperatures climb too high. Fuel vaporization causes rough idling and slow speed running until the car can be driven on the open road, allowing a good flow of air through the grille into the engine bay. To put this problem into perspective, it's only significant for very highly-tuned engines on hot days, in traffic jams or during long periods of slow town driving. The problem

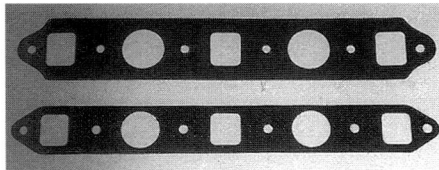

Mini Spares large port manifold gasket (top) alongside standard gasket.

of excess engine compartment heat can be drastically reduced by having louvres cut into the bonnet.

You may be tempted to reduce excessive engine bay heat by wrapping the exhaust manifold with heat insulating tape: this will work, but there is a downside. Insulation causes the pipes to run much hotter than would normally be the case, resulting in substantially reduced manifold life, especially with LCB types. It's certainly the case with Janspeed, and probably with other aftermarket exhaust manifold manufacturers too, that the use of insulation invalidates the manifold manufacturer's warranty. For serious racers the need for regular manifold renewal may be an acceptable price to pay for the successful use of insulation.

Janspeed can supply an extractor manifold and freeflow system for the 1500cc Midget that is the equivalent of the system described here for the A-series engine.

LCB manifold
The Long Centre Branch (LCB) manifold allows exhaust gases to escape from the engine quickly, thereby reducing pumping losses and making the engine more efficient. When selecting a manifold, check the bore size of the pipes and note how smooth the pipe bends are and how much attention has been paid to the joints.

3-into-1 manifold
The alternative to the LCB manifold is the 3-into-1. Oselli is able to supply a 3-into-1 manifold for the Midget/Sprite.

Controlled vortex manifold
A variation on LCB or 3-into-1 manifolds is the incorporation of a controlled vortex (CV) stubs into the manifold.

The CV, or anti-reversion, manifold was developed by Janspeed in conjunction with David Vizard in 1981. At the time,

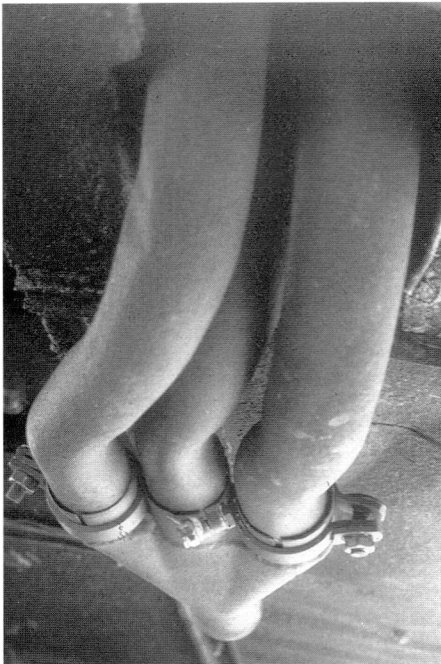

The Janspeed 3-into-1 manifold (header) and collector box.

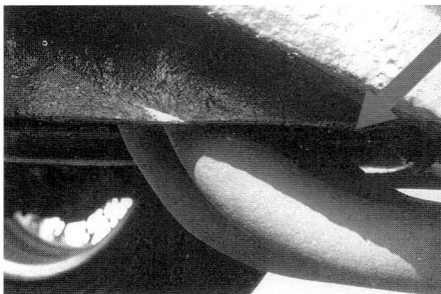

One of the places the car's body needs modification is where the manifold downpipes exit the engine compartment.

Janspeed seriously marketed the CV manifold and it was popular amongst racers. However, the high cost of the manifold made it a less popular conversion for road cars. Also, the real benefits are not realised unless it's used on engines with large carburettor chokes and long period cams: a specification found on only the highest tuned of road cars. Because of the relatively small demand, CV manifolds are now made on a one-off basis and are quite expensive. Janspeed holds the British patent for the CV manifold.

So, what is the CV manifold and how does it work and what are the benefits? The CV manifold is different in that - at the port end of the manifold - it has small cones with a large space around them. From the outside, this part of the manifold just looks slightly thicker than you might expect. With an engine that is running a long duration camshaft, there is a problem with exhaust gas backflowing through the port and into the combustion chamber because the valves are being held open for a long time. The longer valve openings allow for high gas speed and optimum chamber filling at high rpm, but at low rpm can result in lower gas speed, reverse flow and power loss. The cones in the CV manifold hinder the backflow of exhaust gas. To quote Janspeed: "The controlled vortex manifolds are designed for competition cars running wild camshafts with high induction and exhaust gas speeds. The benefits of a controlled vortex manifold are that the engine will gain torque at the bottom to mid rpm range without loss of bhp at the top end. I would, however, stress that the engine will not gain brake horse power at the top end. As a byproduct of increasing torque and hence lowering the usable rpm, you will also find that the car will become far more efficient fuel wise."

The author feels that the CV manifold is very much underrated and offers many benefits. The extra power is very noticeable; his car ,when fitted with a CV manifold, had a noticeable and very pleasing surge in power from about 2700rpm to 4200rpm. The engine pulled much better across the whole rev range and was much smoother through gearchanges, particularly where the ratios were widest, both up and down the box. Changing from an LCB to the 3-into-1 CV manifold produces gains higher up the rev range. For example, real gains were made on what was an already highly-tuned 1312cc engine using a Kent Cams' Megadyne 286 camshaft. In an engine using a longer duration cam, such as the Megadyne 296 Race or 310 Full Race, the advantages of the CV manifold should be even greater.

A radical way of making sure your car's exhaust doesn't ground - Simon Page's turbo Frogeye Sprite.

Ripspeed built a 1600cc A-series engine for a rallycross Mini using a CV manifold and said "it dropped the peak torque figure by 1000rpm," thereby making the engine much more driveable. It's important to remember, too, that it's torque that produces acceleration. The CV manifold will allow the use of a more radical (longer period/increased overlap) camshaft in a road car with minimum loss of tractability.

There are further bonuses in using the Janspeed CV manifold apart from the exhaust benefits. A road car with a competition engine not only has to suffer compromise in cam selection but also in carburettor calibration. On a Weber-engined car the choke (main venturi) will be smaller for maximum flexibility as required for town driving, but larger for maximum power output at high rpm. The most suitable choke size for out-and-out power will, in some cases, give a very flat engine response at 4000rpm, yet a choke size 2mm smaller will be fine in all respects except full power. As the power potential of the engine gets very high the gap between the desirable choke size and the practical choke size increases. With the CV manifold, it's possible to go to the optimum choke size for full power and still retain good bottom end flexibility; in fact, more flexibility than when using small chokes with a conventional LCB manifold.

If you do opt for the Janspeed 3-into-1 CV exhaust manifold your car will probably be required for fitting of the manifold as there isn't a pattern for this combination at the moment. It's also necessary, during fitting, to relieve one or two areas of bodywork to allow adequate clearance for the downpipes; Janspeed undertook this for me, however, and you'll find the necessary information in the body chapter.

EXHAUST SYSTEM

The exhaust system is just as important as the manifold. In some cases more power, without loss of flexibility, can be liberated by a good exhaust system than by a change of manifold. However, it's always preferable to change both at the same time.

The exhaust system on the author's heavily modified Sprite is the Mini Spares RC40, which will flow more cubic feet a minute (cfm) of exhaust gas than any other A-series system. Not only is it very efficient in gas flow and reduction of exhaust back pressure, it's also very efficient in silencing exhaust noise - more efficient than the next best system in cfm terms. The author has found it to be the quietest system he's ever used, and it also allows the best bhp production. Have no fear, though, it does still produce a healthy, rorty exhaust note capable of turning heads!

The only problem when using the RC40 system is that, because it's designed to fit the Mini and Metro, it requires adaptation to fit the Midget/Sprite. The author originally hung his car's RC40 system with the silencer boxes in the original Mini/Metro layout which, on the Midget/Sprite, gives a silencer box each side of the rear axle. You may be tempted to separate the boxes and relocate them. However, Keith Dodd of Mini Spares Ltd. says that the distance between the first and second box was determined by dyno testing and the spacing between boxes can

This is a full race RC40 exhaust system for the Mini.

be critical. An alternative is to use just the rearmost silencer box which is available separately. This was tried on the author's second system and it proved much easier to fit, and not much noisier. Fitting is simply a case of buying some large bore (sized to match the manifold and silencer) stainless steel tube and cutting it to a length that will fit from the manifold collector box pipe to the silencer system. A single straight pipe does the job. Depending which manifold is fitted you may find it necessary to heat and bend the pipe for ease of fitting. When using a LCB or 3-into-1 manifold, the pipe connection to the system may be quite close to the ground and therefore the preferred clamp to use at the join is a ring type as used on

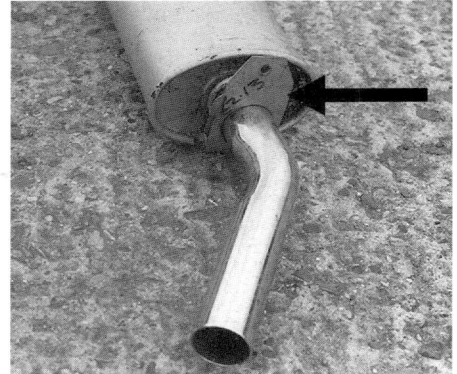

The author hung the single silencer box, via a short flexible strap, on a bracket that mounts on the rear bumper iron. The tailpipe will need shortening.

To suit the Sprite and Midget, the front cylindrical box of the RC40 is cut out and mild steel pipe welded in: this is a good place to fit one of the brackets to hang the system on. Alternatively, the RC40 rear silencer can be purchased as a unit.

Volkswagen cars, as opposed to a conventional U-bracket. The system can be hung on universal flexible mounting brackets of the type available from your local exhaust stockist.

If you have lowered your Midget/Sprite, there can sometimes be a problem with clearance for the silencer box. As a precaution, you may want to weld skid plates to the box and if you are stage, or even pecial stage, rallying your car this will be a necessity.

Chapter 6
Flywheel & Clutch

FLYWHEELS

Each engine size used in the Midget/Sprite employed its own flywheel and the two versions of the 1098cc engines each had their own flywheel.

Irrespective of which engine your car is using, there is an advantage to be had in lightening and balancing the standard flywheel. Lightening the flywheel is beneficial to engine performance because a light flywheel requires less energy to accelerate it. Balancing is beneficial because it reduces vibration and thus stresses on the rest of the engine. There is an old myth that any engine with a lightened flywheel will have a lumpy tickover, but, in the author's experience of four different engine specifications, this is unlikely to be the case.

For the 1500cc engine flywheel balancing is crucial if the engine is going to be worked hard as it has been known for unbalanced flywheels to disintegrate.

Both Motobuild Ltd. and Peter May Engineering Ltd. can lighten and balance flywheels. However, before you despatch your flywheel for lightening, consider whether if, at some point, you might want to use the AP Racing 7.5 inch clutch (more of which later). If you do, ensure that when the flywheel is lightened a chamfer is not machined on the clutch face but that the full width of the face is retained. A typical weight for a lightened flywheel which can be used with the 7.5 inch clutch is 11lb (5kg).

A lightened and balanced standard flywheel is ideal for road use and is suitable for competition use. However, lighter and stronger aftermarket flywheels are available for ultimate competition use.

Steel flywheels

Steel flywheels are readily available for both 1275cc and 1500cc engines. If do you have trouble finding a flywheel to purchase, or require a steel flywheel for a 948cc or 1098cc engine, Farndon Engineering can supply one.

The advantage of a steel flywheel over a lightened standard one is that it is generally lighter and, more importantly, stronger. Ensure your chosen flywheel has a full width face if you are going to use it with an AP Racing 7.5in (190.5mm)

Clutch side of the Farndon Engineering steel flywheel, drilled for 7.5in AP Racing competition clutch.

clutch. The author currently runs a Farndon steel flywheel on his car's engine that, complete with ring gear, weighs in at 9.9lb (4.5kg). Mini Mania in the USA produce a flywheel that weighs 7.9lb (3.6kg). However, note that the Farndon flywheel used by the author is specifically machined to be used with a 7.5in (190.5mm) diameter clutch which pre-cludes machining a larger chamfer on the outer edge of the flywheel, hence the difference in weight.

Cast-iron flywheels

Peter May Engineering can produce a lightened cast-iron flywheel for use with a 7.5in (190.5mm) clutch that is just 0.5lb (0.25kg) heavier than the Farndon steel flywheel - exploding the myth that you can't get the weight of a cast-iron flywheel down. This flywheel is safe up to 7500rpm.

Cast-iron flywheel lightened, balanced and modified to take 7.5 inch (190mm) clutch - all work undertaken by Peter May Engineering Ltd.

Alloy flywheels

An alternative to a lightweight steel flywheel is a Motobuild alloy flywheel for A-series and 1500cc-engined cars. The main body of the flywheel is in alloy, but it has a steel ring gear and steel centre section (which is bolted in) for the clutch face. Both of these steel parts are replaceable and the flywheel itself can be lightened further. This flywheel will not take the larger AP Racing 7.5in (190.5mm) clutch unless you have a special centre machined for it and have the alloy body remachined to match - work Farndon Engineering can undertake for you if you supply a flywheel, or by special order with Motobuild.

The typical weight for an alloy flywheel is 9lb 1oz (4.1kg) complete with ring gear. A point to note about the alloy flywheel is that you must use the bolt locking plate irrespective of what flywheel bolts you use, and whether or not you have thread-locked them. The reason for this is that if the bolts are in direct contact with the flywheel, they will fret into the

Motobuild alloy flywheel for 1098cc engine.

alloy, resulting in an increasingly loose fit and, eventually, a serious problem.

Alloy flywheels dissipate heat better than cast-iron or steel flywheels and are also said to possess better damping characteristics.

Morris Marina flywheels

If you are using the Morris Marina 1.3 A-series or A-Plus-series block, discard the flywheel and use the Midget/Sprite flywheel instead. The Marina flywheel is heavy and very difficult to lighten. The 1275 Midget/Sprite flywheel will fit the Marina crankshaft, but needs modifying to fit an A-Plus-series block crankshaft. The Marina flywheel is incapable of being fitted

to the Midget/Sprite crankshaft so, if you use the Midget/Sprite crankshaft, use the flywheel as well.

Flywheel bolts

Whatever flywheel you use on your car's engine, for the minimal extra expense incurred I recommend you use a high tensile flywheel bolt set as these bolts are far superior to standard bolts. The sets are available from APT.

CLUTCH MODIFICATION OPTIONS

If your car's engine is modified and producing more power than the standard engine, you will almost certainly need to change the standard clutch for an uprated item. AP Racing has a range of specially uprated diaphragm spring (DS) clutch assemblies based on the standard original Borg and Beck unit.

The accompanying chart gives advice on clutch selection.

CLUTCH CONVERSIONS

Uprated 6.5in (165mm) clutch (1275 engine)

The 1275 A-series engine uses a 6.5in (165mm) clutch assembly which is good

Clutch modification options				
Intended usage	Roller clutch release	Competition 6.5in (165mm) plate	Competition 7.5in (190mm) plate & cover	Competition 7.5in (190mm) sintered-metal plate & cover
Mild road	Yes	Yes	No	No
Fast road or mild comp. 100bhp+ at wheels	Yes	Yes	Desirable	No
All-out comp. 100bhp+ at wheels	Yes, where possible	Yes, or sintered-metal	Recommended	Yes, if bhp very high

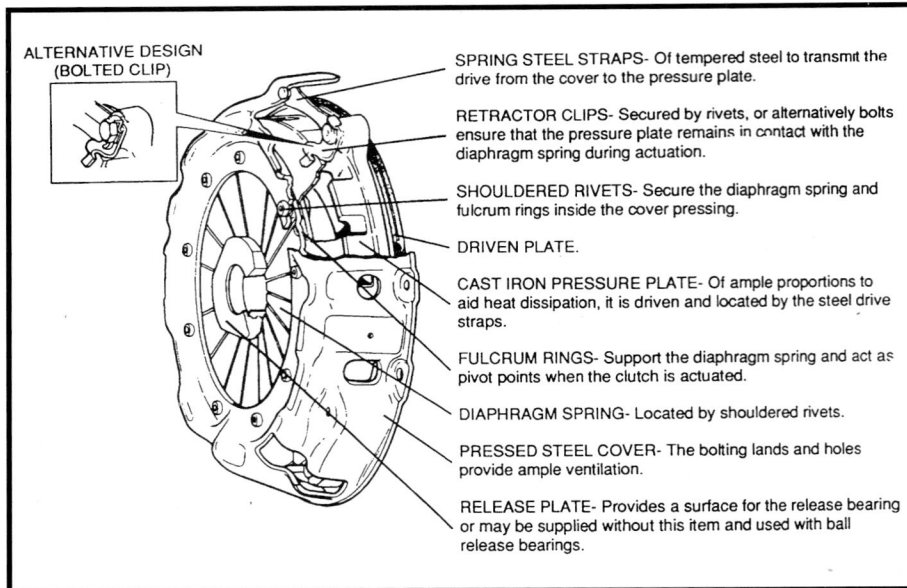

ALTERNATIVE DESIGN (BOLTED CLIP)

SPRING STEEL STRAPS- Of tempered steel to transmit the drive from the cover to the pressure plate.

RETRACTOR CLIPS- Secured by rivets, or alternatively bolts ensure that the pressure plate remains in contact with the diaphragm spring during actuation.

SHOULDERED RIVETS- Secure the diaphragm spring and fulcrum rings inside the cover pressing.

DRIVEN PLATE.

CAST IRON PRESSURE PLATE- Of ample proportions to aid heat dissipation, it is driven and located by the steel drive straps.

FULCRUM RINGS- Support the diaphragm spring and act as pivot points when the clutch is actuated.

DIAPHRAGM SPRING- Located by shouldered rivets.

PRESSED STEEL COVER- The bolting lands and holes provide ample ventilation.

RELEASE PLATE- Provides a surface for the release bearing or may be supplied without this item and used with ball release bearings.

AP Racing DS (diaphragm spring)-type clutch cover for 6.5in or 7.5in diameter clutches. (Courtesy AP Racing).

AP Racing competition 7.5in clutch driven plate and cover bolted to modified flywheel.

Starter motor pinion sleeve which has been relieved to prevent fouling on 7.5in clutch cover.

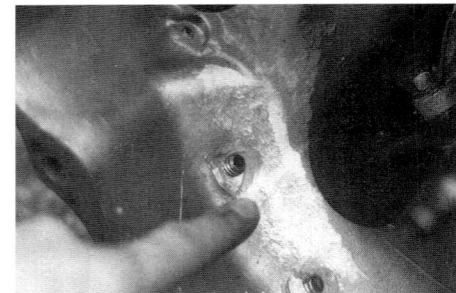

Shiny area is where metal has been ground away to clear 7.5in clutch.

for engine loadings of about 90lbs ft (torque) and weighs in at 5.75lb (2.6kg) for the cover and 1.5lb (0.68kg) for the plate.

For serious road and racing applications, the standard driven plate can be replaced by an uprated item (part number CP2323-6) which is the AP Racing 6.5in (165mm) DS series plate. The author used a 6.5in (165mm) competition clutch disc in his A-series-engined car for several years and only experienced one failure.

There is no longer a 6.5 inches (165mm) competition clutch cover available for the 1275cc A-series engine.

Conversion to 7.5 inch (190.5 mm) clutch (1098 & 1275 engines)

If your car requires a stronger clutch than the 1098cc engine's standard fitment 7.25in (184mm) coil spring clutch or the 1275cc engine's 6.5in (165 mm) diaphragm spring clutch (even in uprated form), it's possible to convert both to a 7.5in (190.5mm) cover and driven plate: the units are available from AP Racing or Peter May Engineering Ltd. There is a

choice of two covers: the AP part number CP2257-5 clutch cover for road use (rated at 140lb ft) and the CP2257-24 clutch cover for competition use (rated at 160lb ft). Note that the torque ratings given are straight torque figures and do not include a safety factor. Note, too, that at the time of going to press, AP Racing was rationalising the CP2257 assemblies and the -5 cover (and all green spring versions), the result of which is likely to be the phasing out of this cover.

The CP2257-24 competition cover is perfectly acceptable for day-to-day driving on the road.

The 7.5in (190.5mm) cover weighs 9.4lb (4.26kg); much heavier than the 6.5in (165mm) cover.

For both 7.5in (190.5mm) covers there is only one driven plate - AP Racing part number CP2257-11 - which weighs 2lb (0.9kg).

An alternative to the CP2257-5 and CP2257-24 covers is the CP2642-1 cover which is a Ford Sierra/Fiesta 7.5in (190.5mm) clutch from AP Racing. This unit is lighter at 6.9lb (3.13kg) but, due to its design, less resistant to heavy down changes (reverse drive loadings). This

clutch would need to be used with a roller release bearing.

To fit a 7.5in (190.5mm) clutch, it's necessary to have the flywheel drilled to take both dowels and bolts to suit the configuration of the larger clutch's pressure plate. It is also necessary to have the flywheel machined flush to take the new, bigger, clutch driven plate. In the UK Peter May Engineering Ltd. can undertake this work. If you're resident outside the UK, your local machine shop should be able to do this work for you if it has a copy of AP installation drawing CP2257 (1275cc engines) or drawing CP2348 (1098cc engines). At the same time as the flywheel is being modified, it's recommended that you have it balanced and lightened as described earlier in this chapter. Have the clutch cover balanced, too.

A problem with this conversion is that the starter motor pinion sleeve may just catch the outer edge of the clutch cover. Before fitting the gearbox, test fit the starter motor as this will highlight any problems. It's preferable that the sleeve be machined in a lathe so that it is waisted at the appropriate point (which depends upon the machining of the individual flywheel). If neither you, or your garage, has the appropriate tool to remove the pinion and sleeve from the starter, an alternative is to use a grinder to grind a rough waisting with the components *in-situ* - no great depth is required.

It is necessary to relieve the gearbox bellhousing to provide sufficient clearance for the bigger clutch. Take care when mating the gearbox to the engine; it may not fit until you've ground out a lot of metal. Places to grind first will be the outermost lug (for the screw that holds the starter motor cover on the gearbox) and the area around the clutch release cylinder bolt holes (right-hand drive cars) - don't forget to shorten the bolts, too! Once you can get the gearbox to fit, turn the engine's crankshaft through at least one full revolution to see whether further relief work needs to be undertaken to achieve full clearance.

When everthing's back together, if you find the clutch travel is right on the limit you'll need to extend the clutch release cylinder pushrod: the procedure is given elsewhere in this chapter.

Uprated clutch (1500 engines)

The 1500 engine can also be fitted with an uprated clutch but requires modification to the flywheel; Motobuild Ltd. or Peter May Engineering Ltd. can supply a suitable clutch assembly. Finally, it is crucial to have the flywheel balanced if the engine is going to be tuned for high rpm operation.

Full-race sintered-metal clutch

Whether you use a 6.5in or 7.5in (165mm or190.5mm) clutch, it's possible to fit a full

Aluminium flywheel with 5.5in Tilton clutch, both available from Speedwell Engineering. (Courtesy Tom Colby, Speedwell Engineering).

Aluminium flywheel with AP Racing 5.5in clutch. (Courtesy Mini Mania).

racing clutch driven plate with a sintered bronze disc (not suitable for road use), associated cover and a special release bearing.

Clutches for Ford-based gearboxes

If you are using the Ford Sierra-based gearbox conversion from the Morris Minor Centre, it can supply you with a suitably-modified Morris Minor flywheel and Sierra clutch that will fit the 1275cc A-series engine. An additional requirement for this configuration is to use a roller release bearing instead of the standard carbon block.

However, if you are already using an AP Racing 7.5in (190.5mm) clutch, you can use your existing flywheel and clutch cover. For the AP Racing 7.5 inch (190.5mm) clutch to work with a Ford-based or Ford Sierra gearbox, your existing CP2257 cover can be used with a 23-spline CP2257-9 driven plate.

If you are using your car for competi-

tion, you can have the added bonus of switching to a cerametallic, rigid, paddle-type driven plate, part number CP2634 to provide the ultimate strength. Using the AP Racing cover with this plate means that you must retain the standard carbon block release bearing.

ROLLER-TYPE RELEASE BEARING

Some racers have found that the standard carbon block clutch release bearings can break. To avoid this, it's advisable to convert the release bearing to a roller type which not only alleviates the problem, but also allows for much quicker and smoother gearchanges. Peter May Engineering Ltd.

Roller clutch release bearing by Peter May Engineering Ltd.

is the only company I know of in the UK which can supply this simple, but useful, modification. The old bearing is replaced by the new bearing and the thrust pad on the clutch cover is discarded, allowing the new bearing to run directly on the clutch fingers; unfortunately, you cannot run this modification on a 7.5in (190.5mm) clutch as the finger spacing is too great.

CLUTCH SLAVE CYLINDER PUSHROD EXTENSION (A-SERIES ENGINES)

On a lightened, or heavily refaced flywheel, used in conjunction with a standard or a 7.5in (190.5mm) clutch, sometimes clutch pedal depression will not result in complete disengagement of the clutch. If you find this to be a problem, an extra long pushrod will be required. The author has used a pushrod extended by 1.5in (38mm) to achieve proper clutch disengagement, though it had to be shortened by approximately 0.25 inches (6mm) when used with a different flywheel on the same car. On this basis, it is best to start with a rod extended by 1.5in (38mm) and then shorten it to the required length for your particular application.

Chapter 7
Gearbox

INTRODUCTION

This chapter covers a number of gearbox-related modifications, ranging from simple gear lever shortening to fitting a replacement 5-speed gearbox. Your choice of modifications will largely depend on how you use your car although, if competition is a consideration, class regulations will also dictate what can be done.

Basically, gearboxes are modified to match engine characteristics to intended use by changing the ratios of some or all of the gears. This can be achieved in one of two ways: fitting replacement gearsets in the existing gearbox casing, or swopping the entire gearbox for a unit with more appropriate characteristics.

ORIGINAL GEARBOXES

Interchangeability

Two different gearboxes were fitted to the A-series-engined Midget/Sprite, both easily identifiable by external appearance as the early box has a smooth outer case, whilst the outer case of the later box is ribbed.

The early-type (smooth outer casing) gearbox will take Morris Minor 1098cc internal components after some machining to get them to fit. It's also possible for both 1275cc and straight-cut gears to be fitted into the early box but, as with fitting 1098cc parts, the work is considerable and includes machining. Alternatively, the whole box can be swopped for the Morris Minor 1098cc unit. If you do choose to use an early box (not recommended), using an alloy backplate (see engine chapter for details) is a worthwhile expense. Straight-cut close ratio gearsets are now only available for the later-type 1275cc gearbox.

Morris Minor 803cc and 948cc gearboxes will also fit the Midget/Sprite, but these have inferior cone-type synchromesh. The 1098cc Morris Minor box internals are not compatible with the later 1098cc Midget/Sprite box except for bearings, baulk rings and layshaft.

The later-type (ribbed) Midget/Sprite gearbox is the strongest and the 1098cc version can be used on 948cc engine cars if the 1098cc engine backplate and clutch assembly are also used. Note that the gearlever and clutch release bearing arm are not interchangeable between 948cc and 1098cc gearboxes.

Whichever A-series gearbox and ratios you do use, there are some uprated parts available from Mini Mania in the USA that are worth fitting. Mini Mania has manufactures layshafts that are precision-made of high quality materials reputed to provide a stronger, straighter, more resilient shaft. The same company also developed a competition baulk ring which is cast in a high quality manganese bronze

Gearbox oil drain plug with magnetic particle trap by Mini Mania. (Courtesy Mini Mania).

alloy and machined and then hand-finished for optimum quality. So, if you have had trouble with baulk rings in a competition box, try the Mini Mania offerings. Also from Mini Mania is a gearbox drain plug that incorporates a magnet.

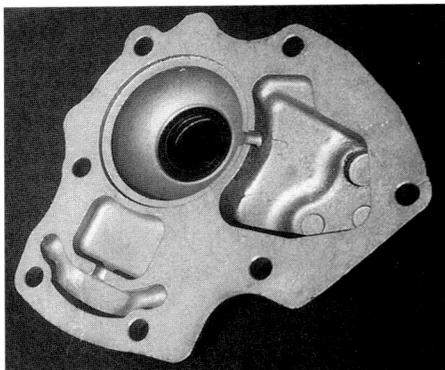

Gearbox end housing with proper oil seal by Mini Mania. *(Courtesy Mini Mania).*

A gearbox modification from Peter May Engineering is a proper lip-type oil seal for the front of the gearbox which will prevent all those annoying gearbox oil leaks. In the USA, Mini Mania produces a special front cover with a proper oil seal.

New ratios and straight-cut gears

If a gear ratio change is contemplated, then only a limited number of choices exist. You can install a set of close ratio straight-cut gears (as close ratio gears are invariably used for competition they are nearly always straight cut; no close ratio

Even straight cut Midget/Sprite gears can break in spectacular fashion - time for a Ford-based BGH gearbox!

helical-cut gears are available for the Midget/Sprite gearbox).

A close ratio gear set is not cheap, but is worth the investment if you're serious about performance. As the fitting of the close ratio gears involves rebuilding the gearbox, it's obviously worth spending the extra money and fully reconditioning the gearbox at the same time. You'll only get what you pay for if you're getting someone else to do the work and your local so-called gearbox 'specialist' might not actually do a good job. It's recommended that you go to a reputable specialist who has real experience of the Midget/Sprite box. The author's recommendation for readers in the UK is the renowned Austin-Healey

specialist Hardy Engineering. Hardy Engineering, Peter May Engineering and Motobuild can all supply gearbox parts.

See the accompanying table for a comparison of gear ratios. Should you wish to work out the mph speed for each ratio, use the simple formula given in the drivetrain chapter.

Webster dog box conversion

A popular gearbox for those racers who can afford it, is a conversion of the later-type (ribbed) Midget/Sprite gearbox to straight-cut close ratios to create a gearbox which has earned the nickname 'the American Hewland.'

The box boasts a much-reduced gearlever action, with almost no chance of missing a change. Due to the nature of its design and, as you might guess from its nickname, you can change ratios to suit the circuit you are racing on and a full range of ratios are available. A real racing box that will last and last, and highly recommended for those racers tied to four forward ratios and the standard gearbox casing.

Shorter gearshift lever

A simple and cheap modification is to

Complete straight cut, close ratio gear set-up for the Midget/Sprite. Noisy but efficient!

Components from the Webster gearbox for the Midget/Sprite.
(Courtesy Tom Colby, Speedwell Engineering).

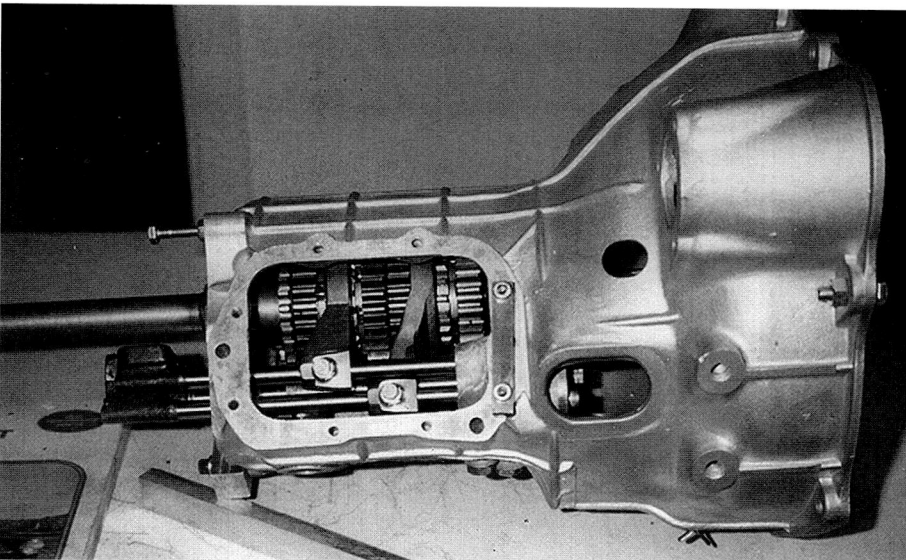

A Webster gearbox part assembled. *(Courtesy Tom Colby, Speedwell Engineering).*

Standard (left), short (centre) and very short gearlevers for the original gearbox. The very short lever made gearchanging a bit on the heavy side but fell readily to hand.

TOYOTA 5-SPEED GEARBOX CONVERSION

An alternative to fitting a straight-cut close ratio gear kit to the original-type gearbox is to substitute a 5-speed Toyota box. This popular conversion kit comes from Dellow Automotive in Australia. Originally available from Rooster Racing, it is now retailed by Minor Mania Ltd. (in the UK) which, as the name suggests, is a company mainly catering for the Morris Minor. Many people have successfully fitted the kit to Minors and Midgets/Sprites.

The conversion is only suitable for A-series-engined Midgets/Sprites and does

shorten the gearshift lever. The standard lever is too long, so shortening it is a good way of improving Midget/Sprite cockpit ergonomics: a short lever aids quick and smooth gearchanges. How much you shorten the lever will depend on your seat position and arm reach, but 1in to 1.5in (25.4 to 38mm) shorter is often ideal, though some racers use a stick a lot shorter than this.

You can approach the job in one of two ways. Either cut the lever in two, cut out the length you require and then weld it back together, or cut the lever to the required height and re-thread the 'new' top of the stick. The choice is dictated only by the tools and equipment at your disposal.

Toyota's 5-speed gearbox squeezed into Simon Page's 'Frogeye' Sprite - note metal braided hose clutch hydraulic line.

Minor Mania supplies Toyota 5-speed conversion, here mounted on a display stand.

involve a slightly different approach than if fitting to the Minor, however, it is quite straightforward if you follow the instructions supplied. The principle advantage of the conversion is that not only do you get the extra overdrive ratio fifth gear, but you also gain synchromesh on first gear. Another advantage of the Toyota box is that it is very strong, and can easily handle the extra torque generated by even turbocharged engine conversions.

The Toyota ratios are fairly similar to those of the Midget/Sprite gearbox though, should you require a set of straight-cut close ratio gears, they are available through Toyota Sport GB and, no doubt, equivalent organizations worldwide. At the time of going to press gear sets were available by special order only from Japan, so allow several months for delivery.

There a couple of disadvantages to this conversion. The Toyota box is heavy at approximately 61lb (28kg), some 13lb (9kg) heavier than the standard 1275 gearbox and that's without taking into account the A-series to Toyota bellhousing weight. The other disadvantage is that the Toyota gearbox is now somewhat scarce and replacement standard gearsets are no longer available.

FORD SIERRA 5-SPEED GEARBOX CONVERSION

As with the Toyota 5-speed conversion, a Morris Minor specialist has led the way with this conversion. The Morris Minor Centre in Birmingham, England, developed a kit for the Morris Minor but has also carried out a conversion to a Midget/Sprite. The conversion has all the benefits of the Toyota conversion, but the gearbox - and more especially gearbox parts - are much more readily available.

The standard Sierra box and special bellhousing weigh in at 66lb (33kg), which is heavier than the Toyota box and bellhousing. However, this need not be a problem as BGH Geartech can convert your gearbox with a special alloy casing made by FlowTech Racing. The FlowTech Racing case when fully built up is 7lb

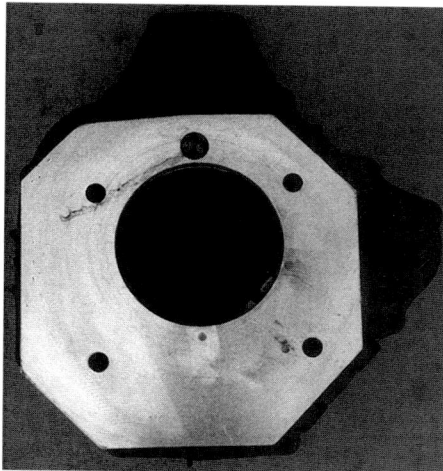

End-on view of the modified bellhousing that allows use of a Ford-based gearbox with the A-series engine.

Interior view of the Ford to A-series modified bellhousing.

(3.5kg) lighter than the standard Sierra box and 3lb (1.5kg) lighter than the Toyota box. Once you have a Sierra box it's also possible to modify it to provide a quicker gearshift and FlowTech Racing can sell you the requisite parts. Another benefit in using the Sierra box is that it is possible to fit any combination of straight-cut close ratio gears to suit. It is also possible to use the 4-speed RS2000 box with a Quaife dog gear conversion.

To learn more about the Ford Sierra conversion the author went to the Morris

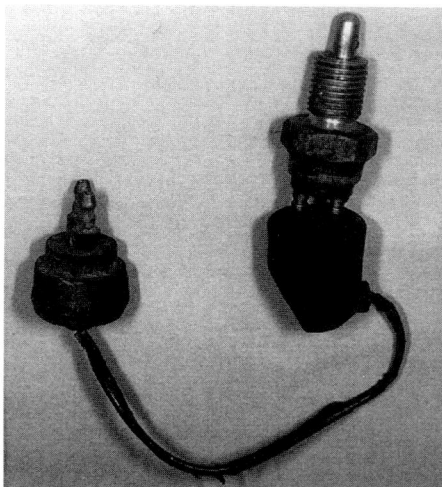

Reversing light switch for Ford-based gearboxes.

Minor centre to have his car converted.

One of the first tasks was to go for a drive in Partner Mike Lennon's Midget, which was absolutely standard apart from the conversion to the Sierra box. The incredibly short gearstick sat further back than with the Midget/Sprite box but still fell easily to hand. The shift was much smoother and faster than the short stick standard Midget/Sprite box in the author's own car. Mike's car was very pleasant to drive with all the benefits of synchro on first and fifth gears and gearbox (standard Sierra) was quiet.

The first stage of the conversion is to remove the engine and gearbox. The gearbox can then be separated from the engine. The standard conversion uses a specially modified flywheel and clutch which is supplied with the kit. An alternative is to use the Midget/Sprite flywheel with a 7.5in (190.5mm) clutch conversion. With this latter choice, a different clutch plate is required in order to fit the Sierra input shaft. Either the early 7.5in (190.5mm) Sierra plate or an AP racing plate can be used.

The crankshaft spigot bearing has to

Modifed release fork attachment on bellhousing of Ford-based conversion.

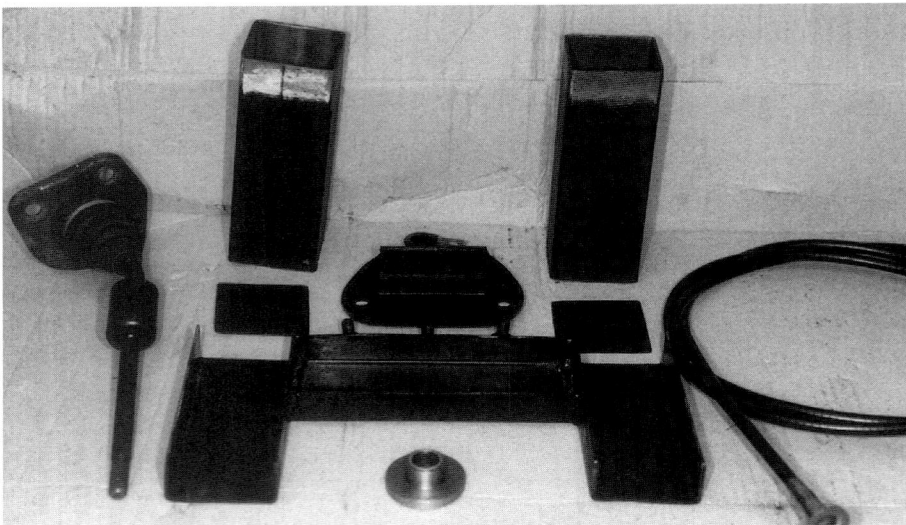

Ford 5-speed gearbox conversion kit parts.

The Ford gearbox conversion requires cutting and welding of transmission tunnel and crossmember. Here are the first cuts in the crossmember. Note that the fuel line has been moved out of the way.

Section of the original crossmember which is cut out: keep it if you think you'll ever refit a standard gearbox.

The gearlever hole in the tunnel has to be repositioned. Despite the lever now being much further back, it feels perfectly placed in use.

be removed and replaced with a special spigot for the conversion - again, supplied with the kit. The Sierra gearbox end piece must be removed and then cut flush and refitted. If used with the AP Racing 7.5in (190.5mm) clutch, the bellhousing must be relieved as necessary to provide clearance. It can then be bolted to the gearbox.

The clutch release fork can be swopped over from the old box to the Sierra hybrid bellhousing, and the box bolted to the bellhousing using the correct Ford gasket.

Before the new gearbox can be fitted to the car some cutting of the bodywork is necessary, the majority of which is to remove a section of the main crossmember that runs across the car. When this is removed and the cuts tidied, a new reinforcing box section is slid into each side of the existing crossmember to strengthen it. On the author's car (which had previously had a new floorpan) there was some resistance to sliding in the box sections and additional cutting and tidying, followed by welding, was necessary to get everything to fit.

The standard Sierra gearlever needs to have the section above the anti-vibration rubber removed; the simplest

Genuine number 5 pool ball is functional and appropriate to ratios of gearbox. Note how transmission tunnel has been repaired.

way to do this, is to heat the stick until the rubber melts and then separate the two sections. Once you have the single short lever the steel collar under the rubber should be ground or cut off. The lever should be heated once more and straightened and then it can be tapped to provide a suitable thread.

A hole must be drilled through the existing crossmember inside the car and also the new H-section member. Once the holes are drilled, the new H-section member can be bolted in. The engine and gearbox can now be fitted to the car. Once everything is in, the transmission tunnel can be cut back to make a suitable aperture for the gearlever. The gearbox will also need to be filled with oil and a new hybrid propshaft fitted.

Two small holes are required in order to fit the speedo cable: one underneath the car in the floorpan to allow entry of the cable and the other in the tunnel section for the cable to exit. Once all this work is done the brake and fuel lines must be relocated. P-clips are recommended for

this job. The lines should be fastened in such a way that they will not chafe anywhere.

SIERRA GEARBOX - BGH CLOSE RATIO CONVERSION

An alternative to expensive and noisy straight-cut racing gears are quieter, helical, close ratio conversions undertaken by BGH Geartech.

BGH Geartech is a specialist company that works mainly with Ford gearboxes. The company markets a range of sporting and racing gearboxes with helical gear cuts but which all feature significant improvements in friction reduction, lubrication, synchromesh quality and life, ratios, breathing and shift quality and strength. The gears used in a BGH box are partly original Ford-manufactured and part BGH designed/Quaife-manufactured.

The author spent the best part of the day with Brian Hill at BGH Geartech and watched him build a sporting close ratio 5-speed box in a FlowTech casing for his

On the left is a BGH special close-ratio helical gear for the Ford-based gearbox.

Various Ford gears. Note gear on bottom left is a BGH special with large roller bearing conversion, as well as special ratio.

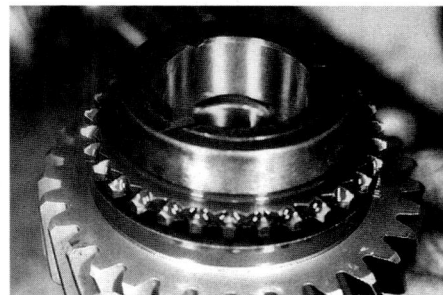

All bearings and synchro hubs are specially modifed in the BGH box.

Even gear selector forks are modifed in the BGH gearbox.

car. It's an understatement to say that Brian is a good engineer; he's a master craftsman. The FlowTech alloy case is not just lighter than the standard case, it's also stronger. Although the average Midget/ Sprite is unlikely to break a standard Sierra case, a turbo or supercharged car just might. The Flow Tech case also has improved breathing capability over the standard case and a breather pipe can be fitted which, if used in a competition car,

Side view of FlowTech Racing Ford Sierra gearbox case which is both stronger and lighter than the standard unit.

Top cover of FlowTech Racing gearbox. Note oil breather outlet.

Ford-based BGH/FlowTech Racing gearbox attached to Sprite engine and ready to go in the car.

BGH Geartech Sierra gearbox ratios					
	1st	2nd	3rd	4th	5th
Standard	3.65	1.97	1.37	1.0	0.82
Long 1st	3.08	1.97	1.37	1.0	0.82
Sporting close	2.92	1.864	1.295	1.0	0.82
Rocket 5	2.58	1.65	1.23	1.0	0.88 or 0.91

will need to feed into a catch tank. Brian hand-finishes each individual piece that goes into the box and just about everything is modified.

A modification to the selector forks makes the change pattern less of an 'H' and more of a lazy 'S' to permit a faster change. Brian's modifications to the lubrication system are vital to the life of the box and something which your average (and even race-built straight-cut) Ford box might not have.

During the build it is possible for a speedometer drive to be selected so that your existing speedo will not need recalibration when the Ford box is fitted. All Brian needs to know is the rear axle ratio your car is using.

When the modified gearbox is fully assembled, it's checked in each gear to ensure everything is spot on. The box is checked for smoothness of change and adjusted as necessary before finally being sealed and stamped with the ratios. Incidentally, there is a choice of fifth gear ratios even with different first to fourth ratios.

CATERHAM 6-SPEED GEARBOX CONVERSION

Although the author doesn't know of anyone who has done this conversion, it seems fairly certain the Caterham 6-speed gearbox will work with the Sierra conversion kit and, therefore, fit the Midget/Sprite.

CALCULATING RPM DROP FOR ANY GIVEN GEAR RATIO

You'll find details of how to calculate vehicle speed for any given tyre/rear axle ratio combination in the chapter on propshaft and rear axle. However, if you want to work out the rpm drop for any given gearchange, the vehicle speed method is hard to calculate and there's an easier way.

To calculate the rpm drop from 1st to 2nd gear, divide the 1st gear ratio of 3.2 by the 2nd gear ratio of 1.916 to get 1.67. If you use this figure to divide the changepoint rpm figure, you get the rpm drop. Here's an example. Changing at 6000rpm from 1st gear to 2nd gear we can divide 6000 by 1.67 to get a rev drop to 3592rpm. If the same gearchange were executed at 5000rpm, the rev drop would be to 2994rpm.

Gear ratios (excluding Ford ratios)					
	1st	2nd	3rd	4th	5th
Frogeye Sprite	3.627	2.374	1.412	1.00	-
Sprite Midget 1275	3.2	1.916	1.357	1.00	-
Sprite Midget special tuning close ratio	2.573	1.722	1.255	1.00	-
Spitfire 1500	3.50	2.16	1.39	1.00	-
Midget 1500	3.4118	2.119	1.433	1.00	-
Midget 1500 close ratio	2.65	1.76	1.25	1.00	-
Toyota Celica 2T	3.587	2.022	1.384	1.00	0.861

The point of these calculations is to illustrate the disadvantage of using standard ratio gears with a race-tuned or seriously modified road engine. For example, a race-tuned engine allied to a standard gearbox will fall off its power curve during the 1st to 2nd gearchange if executed at 5000rpm, but just keeps stays 'on the cam' if the change is executed at 6000rpm. The same gearchange with a close ratio gearbox (1st and 2nd gear ratios of 2.5673 and 1.722) drops the revs to only 3355rpm. On a 1500 Midget a 1st to 2nd gearchange has a rev drop from 5000 to 3105rpm. The same 1st to 2nd gear change on the Toyota gearbox drops the rpm to a low 2820rpm.

From these sample calculations you can see that having a close ratio gearbox is important if you have a race or tuned road engine with a narrower power band that starts higher up the rev range than about 3000rpm. You can also see that the Toyota box has its limitations when used with a high performance engine and that the 1500 Midget engine and gearbox are a pretty poor combination for tuning.

OVERDRIVE CONVERSIONS

1500 Midget

For owners of 1500 Midgets it is possible to fit an overdrive unit to the existing gearbox. Although it's possible to do this yourself, it's easier to visit the Nottingham MG Centre, which specialises in this conversion. The Centre can either sell you a kit, with or without a clutch, or carry out the work for you.

An overdrive is a sun and planet gear mechanism electronically switched in and out to provide additional gear ratios - usually on both third and fourth gears. The extra ratios are activated by a switch and you do not have to use the clutch for engagement. Such a gearbox, which is really a gearbox and overdrive unit combined, was fitted to the 1500 Triumph Spitfire but never to the 1500 Midget. The overdrive ratio used on the Triumph box is 0.797 to 1 which, in use, reduces rpm by approximately 20 per cent.

Fitting entails removal of the existing engine and gearbox unit, then, in order to get the overdrive box to fit, it's necessary to cut out the floor crossmember and weld in a special replacement item to accommodate the new unit. There is a template to ensure that the cutting is accurate. The tunnel will need to be widened slightly and this can be done with a jack and blocks of wood - about an 1in (25mm) each side will do. A test fit is recommended to ensure everything fits okay before proceeding to remove the interior carpets ready to weld in the new crossmember.

The propshaft will not usually need modifying, though any variation in suspension from standard might result in the need for it to be shortened slightly. The overdrive switch gear will need to be wired into the Midget loom after comparing the 1500 Midget and 1500 Spitfire wiring diagrams to ensure the job is done correctly. The speedo from the Midget will need recalibration and will also require a special cable to be made up. For details of both these jobs, see the instruments chapter.

A-series-engined cars

For A-series-engined cars one way to fit an overdrive unit is to install a Morris Marina bellhousing mated to a Triumph Dolomite 1850HL overdrive gearbox. This will involve modifying the transmission tunnel to get everything to fit, as well as modifying several other parts such as clutch hydraulics and starter motor fittings. The disadvantage of this conversion is that it is extremely heavy as both the Marina bellhousing and 1850HL gearbox are made of cast iron. One owner used the Dolomite alloy bellhousing instead of the heavy cast iron Marina item by fitting a specially made alloy backplate to mate the bellhousing to the Midget/Sprite block. The benefits of the conversion are, of course, a fully synchromesh box with overdrive on 3rd and 4th gears.

Another way to fit an overdrive unit is to mate a specially modified overdrive unit to the back of the existing gearbox. The unit will be a Laycock de Normanville J-type model with an output flange at each end. With suitable bracketing and switching, the 4-speed Midget/Sprite becomes a 6-speed car. The work involved in this modification is, however, considerable.

GEARBOX (TRANSMISSION) LUBRICANTS

For competition and high performance road use, it's strongly recommended that you use synthetic oils. The author has found Mobil products work well: 5/50 Mobil One for non-1500cc gearboxes and SAE Mobil gear oil for 1500cc car gearboxes and 5-speed gearboxes.

Chapter 8
Propshaft & Rear Axle

MODIFICATION OPTIONS

The drivetrain can be strengthened and, of course, the crownwheel and pinion unit offers the possibility of a final drive ratio change. Lastly, a limited slip differential can improve traction hugely.

The accompanying chart shows which modifications are desirable in relation to intended usage.

PROPSHAFT BALANCING

Balancing the propshaft may sound like a luxury, but experience has strongly indicated that racing or use of a low final drive ratio will quickly bring to light any propshaft imbalance problems. Since balancing is a specialist job, the author took his car's propshaft to a specialist company - Reco-Prop - for balancing and a full overhaul. The propshaft showed excessive wear and one of the universal joints (UJ) showed signs of seizing, although in everyday use nothing had felt amiss. The UJs were removed and new cups and bearings fitted. Rust was removed from the shaft before balancing and, once balanced, the shaft was even given a fresh coat of paint. When you refit the shaft, make sure you use new self-locking nuts and new bolts.

If you are converting your car to a 5-speed gearbox, you'll need to modify the propshaft. Usually this involves mating the

Propellor shaft being dynamically balanced at Reco Prop. (Courtesy Reco Prop).

Modification options			
Intended usage	Balance propshaft	Limited slip differential	Competition halfshafts
Std/mild road	Yes	Desirable	No
Fast road	Yes	Very desirable	Only for high output engines
Competition	Yes	If class regulations permit	Yes

Propshaft and UJs completely dismantled.
Note dry yoke. (Courtesy Reco Prop).

Rear axle ratios		
Ratio	Number of teeth	Late- or early-type differential carrier*
3.55	11/39	Fits late-type only (special, and very rare)
3.727	11/41	Available for early- and late-types
3.9	10/39	Available for early- and late-types
4.222	9/38	Available for early- and late-types
4.555	9/41	Available for early- and late-types
4.875	8/39	Available for early-type only (A35 van)
5.375	8/43	Available for early-type only (Morris GPO van)

Cars from H-AN7-24732 and G-AN2-16184 onwards, will only take late-type ratios. Axles with late- type differential carrier assemblies can be identified by having the oil filler plug in the differential housing as opposed the axle case.

Note: although the same ratio may be available for early and late type differential carriers, they are not always interchangeable, even with modification. The differences are mainly in the pinion, spacer and bearing.

front yoke and part of the shaft from the donor car to the rear section of the original Midget/Sprite shaft. However, in the case of any gearbox conversion, have a new shaft made up with the correct yokes at each end.

DIFFERENTIAL, CROWNWHEEL AND PINION

Moving on down the driveline we come to the differential, crownwheel and pinion. For ease of understanding it's best to view this group of components as two separate assemblies: 1) the crownwheel and pinion unit which turns the drive through 90 degrees and provides a drive ratio between pinion and crownwheel and 2) the differential which allows the two rear

Standard-type differential unit in early-type carrier.

wheels to turn at different speeds.

One of the most important considerations when it comes to the back axle is choosing a final drive ratio to suit the car's intended use. Your choice will also be influenced by the gearbox ratios used, be they standard, close ratio, standard 5-speed gears or close ratio 5-speed.

There are three commonly available final drive ratios for the Midget/Sprite: 3.727, 3.9 and 4.222. Lesser known, and much lower ratios, are available from the parts bins of the Morris quarter-ton van (4.555), A35 van (4.875) and the Morris GPO van (5.375). Additionally, it is always possible to have a set specially cut by an expert gear cutter like Schofield & Samson Ltd.

Differential carriers come in early or late types and not all ratios are interchangeable, as the accompanying chart illustrates.

It's possible to calculate the potential speed of the car with different gearings by using rear axle and gearbox ratios to arrive at the correct figure for any given engine speed. Here is the simple formula to calculate speed.

$$\frac{60,000}{RAR \& WRPM} = \text{mph per 1000rpm in top gear.}$$

("RAR" = rear axle ratio, "WRPM" = wheel revs per mile)

Calculate miles per hour in intermediate gears using this formula -

$$\frac{\text{Top gear mph per 1000 rpm}}{\text{Gear ratio}}$$

To get the wheel revs per mile figure you will need to contact your tyre dealer or tyre manufacturer. However, the following are examples of some approximate wrpm for a selection of tyres:

145 x 13 = 936wrpm
155 x 13 = 914wrpm
175 x 13 = 908wrpm

Having obtained the necessary values, it is now a simple matter to complete the speed calculation, as follows.

Assuming our test car has a 3.9 rear axle ratio and runs on 145 x 13 tyres, we can work out that 60,000 divided by 3.9 x 936 = 16.44mph per 1000 engine rpm in top gear. We want the figure for, say, 5000rpm in top gear, so it's a simple case of multiplying our 1000rpm figure by five, ie 5 x 16.44mph = 82mph.

To work out speeds in intermediate gears, simply divide the top gear speed at 1000rpm by the relevant intermediate gear ratio. For example, using the 16.44mph top gear at 1000rpm figure divided by the standard third gear ratio of 1.412:1 (16.44 divided by 1.412 = 11.64mph per 1000rpm) we get the 1000rpm speed of 11.64mph for third gear. Multiply this figure appropriately to get the car's speed at any engine speed: *ie* 5 x 11.64 = 58mph (at 5000rpm).

If you don't know the ratio of your rear axle all you have to do is count the number of teeth on the crownwheel and pinion, then divide the number for the crownwheel by the number for the pinion. For example, 41 teeth on the crownwheel divided by 11 teeth on the pinion is 3.727. What the ratio actually means is the number of turns of the propshaft to the number of turns of the axle halfshafts and, ultimately, the wheels. An example is the standard 3.9 rear axle ratio, which means that the propshaft turns 3.9 times for each revolution of the halfshafts/wheels. From this, you can see why numerically smaller numbers for the rear axle ratio actually produce higher gearing and vice versa.

LIMITED SLIP DIFFERENTIALS

Competition type differentials, which replace the standard differential unit, can be of the limited slip or Torsen automatic torque biasing type. Both types are designed to improve traction by feeding power to both wheels at all times; the wheel with best grip getting the most. The benefits can be felt not only during race-type standing starts, but also when attempting to put maximum power down hard at the exit of a corner, especially in the wet.

The most common type of limited slip differential (LSD) is the clutch plate type where the friction of the clutch plates ensures that drive is never totally lost to either wheel. Motobuild Ltd. retail this type of differential. In the USA, Mini Mania retail the Tran-X limited slip differential which is a plate-type unit so, like most plate-types, it is possible to change the static settings. In fact, Mini

Clutch plate type LSD, less cover - this is actually a Jaguar unit.

Quaife Torsen differential in early-type carrier with 3.9 ratio crownwheel and pinion.

Mania offers the differential with three different settings, which can be individually selected by using plates of different

Motobuild-supplied LSD for Midget/Sprite.

thickness. Well-known US Sprite/Midget racer, Joe Huffaker, uses this type of differential.

The alternative to the conventional LSD is the Torsen-type automatic torque biasing differential, which employs helical gears and which, when one wheel spins, locks and transmits power equally across the differential. Quaife manufactures and retails the Torsen principle differential and it's available direct from that company in the UK and, in the United States, from retailers such as Victoria British Ltd.

In use each of these limited slip differentials will make a substantial

Standard (with carrier bearings) and Quaife differentials side-by-side.

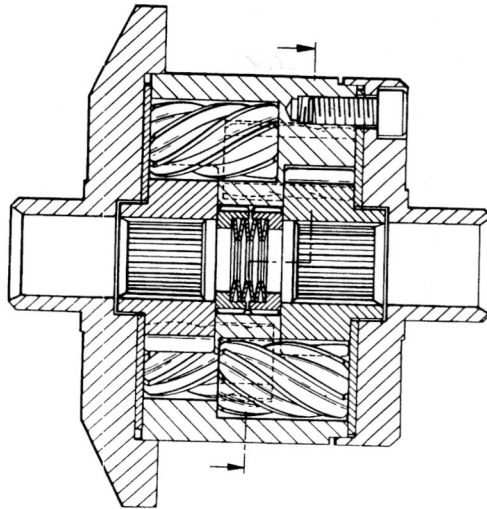

Here is a cross-section of the assembled Quaife LSD ...

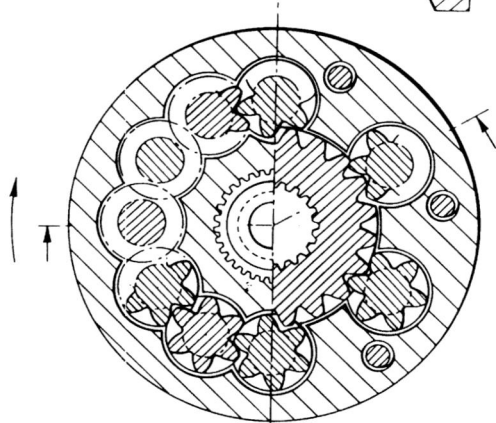

... and here is another from a different axis. This cross-section is split level (see arrows in previous drawing) and shows the relationship between the two sets of sun and planet gears - one set for each halfshaft. (Courtesy Quaife).

new bearings; however, the second major problem may be experienced at this stage, the crownwheel and diff may be too large for the carrier. If so, the only way to get the non-standard crownwheel to fit is to use a different, later carrier or machine the existing carrier.

If you are using a Quaife differential with 3.9, 3.7, or 4.2 ratios, it's recommended that you use a later carrier.

Although it is not deemed necessary to run-in the new differential, it's a good idea to treat it gently for the first few miles and to carry out a precautionary 500 mile (800km) first oil change. The use of the synthetic gear oil, such as Mobilube SHC (SAE 75W-90), is recommended; expensive but worth it.

COMPETITION HALFSHAFTS

If you are using your car for competition it's possible to get uprated halfshafts (Rover part number C-BTA-939 for wire wheels and C-BTA-940 for normal wheels). If you are unable to obtain

Differential end of two halfshafts (for wire wheeled car). The uppermost, twisted, shaft has suffered from a combination of brute power and Quaife traction on Goodyear NCT tyres. Standing starts were amazingly quick, but more than the standard shaft could cope with.

improvement to the handling of the car, whether in day-to-day driving (especially in the wet) or competition work. I understand from some racers that the Torsen Quaife differential is much more user-friendly than the clutch plate type LSD and less likely to let the car snake during a racing start.

Having decided to fit an LSD unit, and made your choice of type, it's time to get it fitted. You can, of course, do the job yourself, but it may be worth letting the manufacturer do the job for you. R. T. Quaife, for example, prefers that differential units which are to be fitted with the Torsen LSD are delivered complete, rather than in a stripped-down state.

CHANGING REAR AXLE RATIOS

On the face of it changing the final drive ratio should be straightforward but, in practice, this may not be the case because, as mentioned earlier, not all carriers and casings accept all ratios. It may not be apparent until the differential carrier is stripped whether the new ratios are compatible; if they are different, you'll find that both spacers and thrust washers for the pinion will need to be machined to fit or new parts turned from steel bar.

Once the pinion has been shimmed and fitted with new bearings and oil seal, the crownwheel and differential can be offered up to the carrier complete with competition halfshafts using these numbers, it's possible to have shafts for wire wheel cars specially made by R T Quaife Engineering and Peter May Engineering can supply competition shafts for steel- or alloy-wheeled cars. Motobuild can also supply competition halfshafts. Only cars with over a hundred brake horsepower at the wheels need competition halfshafts, but reduce this figure to 80bhp at the wheels if you're using slick racing tyres.

DOUBLE BEARING REAR WHEEL HUBS

Double bearings help eliminate axle flex and therefore reduce the likelihood of halfshaft breakage. Note that bolt-on wheel-type hubs are different to wire wheel-type hubs. When fitting the double bearing hub kit, buy new studs as it saves a lot of aggravation later.

The double bearing hub kit (which

Double bearing rear hubs from the USA. (Courtesy Tom Colby, Speedwell Engineering).

really will eliminate any trace of rear wheel steer) is available in the USA from Tom Colby's Speedwell Engineering or Mini Mania and, in the UK, from Mini Spares and, possibly, Frontline Spridget Ltd., too.

At the time of writing, double bearing hub kits are not available for wire-wheeled cars.

Chapter 9
Suspension & Steering

FRONT SUSPENSION MODIFICATION OPTIONS

Before modifying your car's suspension, decide how much you can afford to spend. Think, too, about what you want from the suspension: is it for fast road or motorsport use, or a compromise between the two? Also, it's most important to match front and rear suspension modifications so that they work in harmony: a full race front end with a standard rear end will not produce good results.

It is wise to modify the car's suspension a stage at a time, and to experience and consider what affect each change produces. This progressive approach will help you to tailor your car's handling characteristics to your own preferences.

If parts of your car's suspension are worn it's unlikely to give good results, no

MGB lever arm damper (disabled) and telescopic shock absorber fitted to Graeme Adams' racing Midget as suspension top link.

Ingeniously, an adjustable tie rod has been added to the damper arm/top link of this racing Spridget. The tie rod triangulates the damper arm, effectively turning it into a wishbone.

Front suspension modification options				
Intended usage	Spring rate	Spring free length	Roll bar diameter	Shock absorber (damper) stiffness
Standard/ mild performance road	275lbf/in (3.169kgf/m)	9.4 to10.2in (259mm)	1/2in (12.7mm)	Std
	280lbf/in (3.226kgf/m)	9.85in (250mm)	9/16in (14.2mm)	Std
	329lbf/in (3.686kgf/m)	8.27in (210mm)	5/8 or 11/16in (15.8 or 17.4mm)	+30%
Fast road use	340lbf/in (3.917kgf/m)	8.74in (222m)	11/16 or 3/4in (17.4 or 19mm)	+30 to 50%
	350lbf/in	9.0in (228.6)	11/16, 3/4 or 7/8in (17.4, 19 or 22.2mm)	As required
Competition use	360lbf/in (4.147kgf/m)	8.03in (204mm)	3/4 or 7/8in (19 or 22.2mm)	As required
	370lbf/in	7.25in (184.1mm)	3/4 or 7/8in (19 or 22.2mm)	As required
	400lbf/in	7.28in (184.9mm)	3/4 or 7/8in (19 or 22.2mm)	As required
	430lbf/in	7.32in (185.9mm)	3/4 or 7/8in (19 or 22.2mm)	As required
	510lbf/in	7.25in (184.1mm)	3/4 or 7/8in (19 or 22.2mm)	As required
	520lbf/in	7.25in (184.1mm)	3/4 or 7/8in (19 or 22.2mm)	As required

car will be dictated not only by what you can afford and your intended application, but also, most importantly, by your own driving preferences. Nevertheless, unless you are a serious motor racing contender searching for 10/10ths performance, you'll find it relatively easy to substantially improve the handling of your car.

The accompanying chart gives guidelines on how to make a good choice of matching front suspension components.

REAR SUSPENSION MODIFICATION OPTIONS

The rear suspension is straightforward in design and consists of a live axle, leaf springs and lever arm dampers (shock absorbers). Early cars have quarter-elliptic leaf springs, while later models use half-elliptic. The 1500cc Midget has six leaves to its springs as opposed to the five found on half-elliptic cars.

If your car handles badly, it's not due to poor design as, when in good condition, the standard system is fine for cars with standard performance; it's more likely the suspension is simply worn after many years of service. The most cost-effective way to modify the suspension is to replace standard parts with modified ones during a

Simon Page's fabricated rear suspension with vertical coil over shock absorber suspension units.

matter how you set it up. Therefore, ensure that all the suspension components are in good, serviceable order before you start fitting high performance parts to the system. Standard overhaul of the suspension is described in workshop manuals and another useful source of information is the Ron Hopkinson Spridget catalogue, it has well-written hints and tips-style guidance to suspension rebuilds.

When modifying suspension it is important to consider the weight of the modifications. The suspension compo-

nents form part of the unsprung weight of the car and reductions here make the job of the spring and shock absorber easier, contributing to improved ride and traction. Unsprung weight is comprised of the combined weights of wheels, tyres, brakes, hubs and half the weight of the springs and shocks.

This chapter details various Midget/ Sprite suspension options but, really, only scratches the surface of what is a vast subject - and one with many variables. Ultimately, the suspension set-up for your

Rear suspension modification options					
Intended usage	Nylatron bushing	Polyurethane bushing	Telescopic dampers	Reduce ride height	Anti-tramp bars
Standard/mild road	No	Yes	Recommended	No	No
Fast road	Optional	Yes	Recommended	1 inch (25.4mm)	Yes (if over 85bhp at rear wheels)
Competition use	Yes	Yes	Recommended	2 inches (51mm), plus	Yes

Fully modified racing Midget rear suspension features coil springs and much, much more. Not for the faint-hearted!

full suspension rebuild.

When working on the car's rear suspension, the best and safest way to tackle a rebuild or fitment of modified parts is to support the car with proper axle stands placed just forward of the front spring hangers. Also, be sure to chock the front wheels.

To help you decide what is best for your car, the accompanying chart details various options.

It is strongly recommended that, if possible, you only make one change at a time to the suspension so as to better evaluate the resultant change in the car's handling.

OVERSTEER & UNDERSTEER

Whenever a car turns, the change in direction of its mass creates a sideways pressure on the tyres. This results in the tyre running with a 'slip angle' (the angle between the direction in which it is pointing and the direction in which it is being forced to travel by the inertia of the car to which it is attached). The front and rear tyres rarely run at the same angle when the car is cornering. When the front tyres run the greatest slip angle the effect is known as 'understeer' and is usually felt by the driver as the front end of the car wanting to run slightly wide of the intended cornering line. If the rear slip angle is the greater, the condition is known as 'oversteer' and the driver usually feels the rear end of the car wanting to run slightly wide of the corner. The Midget/Sprite is generally an oversteering car due to its front engine/rear drive layout.

Understeer & oversteer correction

Reduce understeer
Fit softer anti-roll bar (front)
Fit softer front springs
Use more negative camber

Reduce oversteer
Fit stiffer anti-roll bar (front)
Fit stiffer front springs

Careful selection of suspension parts and tyre width can, however, change the standard characteristics of the car.

To assist in resolving understeer and oversteer problems consult the accompanying chart. Remember, though, you must first establish whether the handling problem you are trying to resolve - be it understeer or oversteer - lies with the front or rear suspension.

FRONT SPRING MODIFICATIONS

One of the key components in deciding your car's ride characteristics are the road springs. Road springs govern ride stiffness and also dictate ride height. Your choice of road spring will be based on the ride height you are looking for and how stiff you can go for your desired application. For instance, the smoother your intended road or track surface the lower and harder you can go, if you so choose.

Several retail outlets and manufacturers will sell you road or fast road springs in a range of heights. However, quite often, these same people don't really know what they are selling or what results you'll get. The author was once sold a set of lowered, and very stiff, springs for his Sprite and found that not only were the springs stiffer than nearly every racing Sprite's, but lower too! Yet, he still found them acceptable. Always check to be sure you know what you are buying and, more importantly, how the springs compare with the standard items.

As a simple guide, hard springs will give improved cornering and straight line stability by reducing body roll. Additionally, stiffer springs reduce weight transfer from rear to front under heavy braking which helps alleviate rear wheel lock-up. Reduced height springs will move the car's centre of gravity closer to the road or race track, which has the effect of reducing loading on the car when moving, be it side-to-side or end-to-end. The result is greater stability under all conditions.

The spring rate for most Midgets and Sprites is 271lbf/in (3.127kgf/m) though late 1500 Midgets use a rate of 275lbf/in (3.169kgf/m). There are four different standard spring heights (free lengths), as detailed in the chart earlier in this chapter. When you see a spring advertised as one inch lower, ask yourself this: one inch lower than which of the four standard ride heights? Ask the retailer what the exact measured height is. If the car is to be used on the public highway, generally, it's not a good idea for ride height to reduced by much more than 1in (25mm) for any model except 1500 Midgets, where an extra half-inch (12.5mm) lower is acceptable.

Some spring retailers appear unable to quote the spring poundages/rates or heights. Aftermarket spring poundages/rates for the Midget/Sprite are usually less than 450lbf/in (5.184 kgf/m) and springs taller in free length than 7.28 inches (185mm). My experience has been that you can get good advice on spring selection from Peter May Engineering. Ltd, a company which can supply, to order, any spring specification you request, as can AVO UK Chassis Dynamics and Motobuild while, in the USA, Mini Mania and Speedwell offer the same services.

As a starting point and for road use, springs which are one inch lower than standard and rated at around 329lbf/in (3.687 kgf/m) are recommended. For motorsport, try shorter and harder examples. Be careful how low you go and check, even with one inch lower springs, that there is enough ground clearance for the exhaust system which is generally the lowest point on the car.

A change of spring is not the only

Spring pan lowering kit.
(Courtesy Victoria British Ltd.).

Huffaker Racing front suspension. (Courtesy Mini Mania).

way to lower the suspension, Motobuild Ltd and, in the USA, Victoria British Ltd, retail a spring pan lowering kit which lowers the spring in relation to the wishbone (A-arm), a method which can be used to lower the car in relation to any height of spring. This gives a much reduced ride height without the usually associated stiffness of shorter competition springs.

One important factor you'll need to take account of when using the spring selection chart is any variation from a standard car's weight. If you have fitted an aluminium engine backplate and a lightweight fibreglass front to your car, it will be considerably lighter than a standard car and will therefore not need such a high spring rate. Assuming you are using the car for fast road use, you might find that the car is better with 329lbf/in (3.686 kgf/m) than 340lbf/in (3.917 kgf/m) springs, for instance.

Conversely, if you have fitted a five-speed conversion to your car and/or a roll over bar, you'll have a heavier than standard car and may need to go above my recommended spring rates to achieve satisfactory handling, maybe even using

360lbf/in (4.147 kgf/m) springs on a fast road car.

At the end of the day, only you can decide what is an acceptable ride and how hard you want to go on spring rates.

HUFFAKER RACING SUSPENSION KIT

Huffaker Racing has created a very special suspension system for the Midget/Sprite. Fitted to Production National Championship Runoff cars in the USA, it was, arguably, the reason the cars won. British racers note that a win at this level of racing in the USA is truly outstanding; it's hard even to get a place on the grid. That's not to say that US racers would beat British racers hands down, but the author, for one, would like to see a US v British Midget/Sprite race. MK Parts in Germany retail Huffaker Racing parts and Mini Spares in the UK can supply them, too.

Front suspension
As you will see from the accompanying photos, the Huffaker front suspension is

rod ends and fasteners, Watts-linkage integral frame and, finally, a modified differential carrier (housing).

Fitting the kit entails welding the Watts link frame, shock absorbers brackets and trailing link brackets to the car.

RON HOPKINSON HANDLING KIT (FRONT & REAR SUSPENSION)

The author was one of the first people in the UK to get hold of this kit and, since his car's suspension is already highly modified, he fitted the kit to Phil Bollen's Midget, which already had uprated Armstrong dampers, slightly stiffer rear springs and an 11/16in (17mm) anti-roll bar.

Because of these existing modifications the improvements brought by the kit are less noticeable than they would be with an otherwise standard car, but were

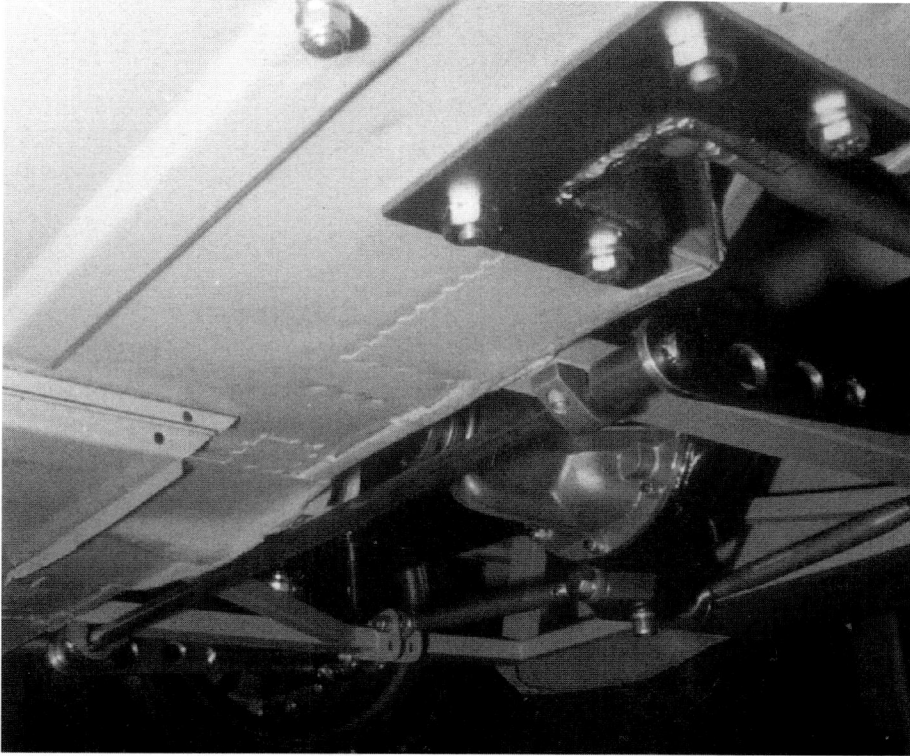

The radical Huffaker Racing rear suspension kit. (Courtesy Mini Mania).

very special indeed, consisting of fabricated wishbones with race-bred geometry. Everything is adjustable, from camber and castor to roll centre, damping and ride height. If all those adjustments sound like a nightmare, don't worry, as initial settings are included with each kit. Some fabrication is required when installing the kit for the first time and a good race preparation specialist should be capable of this.

The Huffaker front suspension kit comprises Carrera adjustable shock absorbers, springs, shock absorber mounting towers, shock absorber tower cross brace, upper and lower wishbones, rod ends and locknuts, fasteners and washers, modified stub axles, modified steering arms, anti-roll bar with mountings and linkages, modified steering rack, tabs, brackets, gussets and pick-up points.

Fitting involves the addition of lower wishbone attachment points. An upper trailing arm attachment point also needs to

be fitted. The standard spring top platform outer edge has to be cut away to allow clearance for the new coil-over spring/shock absorber assemblies. The inner wing panel (if you still have one) needs modifying, too. You'll have to drill holes for the shock absorber tower attachment. After all the welding has been completed the bottom wishbone spacers are made, followed by checks for clearance and adjustment of the anti-roll bars and mountings.

Rear suspension
This competition rear suspension set-up is available from Huffaker Racing and Mini Mania in the USA, Mini Spares in the UK and MK parts in Germany. This is a race-proven three-link system which includes the following components: adjustable three-link trailing arms, adjustable horizontal Watts-linkage, adjustable anti-roll (sway) bar and links, adjustable shock absorbers with upper and lower pickups,

Ron Hopkinson handling kit.

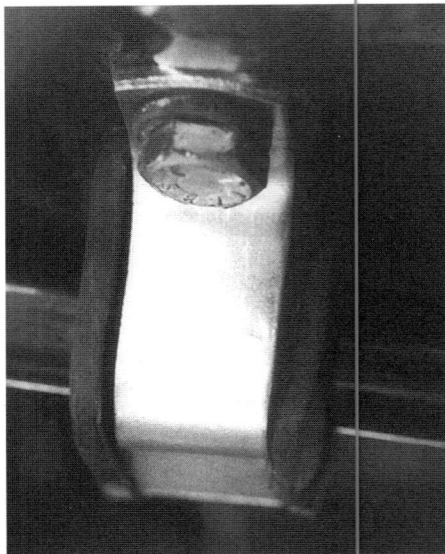

Rubber anti-roll bar mounting of Ron Hopkinson handling kit.

still very worthwhile. The kit was fitted in two stages to evaluate the rear and front components separately. With just the rear components fitted a test drive showed an enormous improvement in handling; with the rest of the kit fitted the whole character of the car had been transformed and it displayed vastly improved handling.

The kit was developed using the latest computer technology and comprises spring clamps with nuts, washers and bolts to fasten them, uprated rubber bearings/ bushes, three-quarter anti-roll bar with bearings/bushes, brackets and bolts.

The kit claims to eliminate low speed, tight corner understeer and higher speed oversteer without sacrificing ride quality and the claim is justifiable. Certainly, roll is much reduced, improving overall stability. The axle clamps also help reduce axle tramp on powerful cars where this is a problem.

Fitting instructions - text and a diagram - are included with the kit. The spring clamps are fitted in different positions, depending on whether your car is rubber- or chrome-bumpered. Fitting is straightforward, although it may be necessary to open out one or two of the holes on the spring clamps where con- struction of the clamp half (usually the top one) during welding has slightly obscured the hole. Aside from this small point, fitting is as per the instructions.

To summarise, this is an excellent, value-for-money, easy-to-fit kit, which will transform the handling of any Midget/ Sprite. It's worth fitting to the car which has no other suspension modifications and

to the car which already has uprated springs and dampers.

REAR SPRING MODIFICATIONS

Motobuild is a good company from which to purchase uprated rear springs and Rae Davis of Motobuild can offer good advice on the best springs for your particular car. Standard half-elliptic leaf springs were made in three different spring rates: 75lbs/ in for 1098 cars, 80lbs/in for 1275 cars and 86lbs/in for 1500 cars. The 1500 leaf springs had six leaves whilst both the earlier cars had five leaves in the springs. The free camber for each spring is, respectively, 4.437in, 4.72in and 5.58in (113mm, 120mm, 142mm). Springs that produce a lowered ride height are available in both 1in (25mm) and 2in (50mm) reduction for 1275 and 1500 cars. Cars that are 1098-engined but have half-elliptic springs, can use the 1275 range of standard or lowered springs, too.

Some springs are uprated by the use of extra leaves, although it's not so much how many leaves there are but rather the lengths of the top leaves that stiffens the suspension. Of course, race cars which have been substantially lightened may not need stiffer springs at all and, generally speaking, Midgets/Sprites do not require stiffer rear springs. If, however, you do want a stiffer spring, Motobuild can have one made for you. At the time of going to press, MK parts in Germany sent me details of a lightweight and 1in (25mm) lowered spring which sounds like it could be a good choice.

Whereas it is usual to go stiffer on springs for improved fast driving handling, standard quarter-elliptic cars are oversprung rather than undersprung. Because of this, Tom Colby's Speedwell Engineering supplies a spring which is 20 per cent softer than the standard item. Also from the USA is a very special offset quarter-elliptic spring from Faspec. These quarter-elliptic springs are specially offset to allow the fitting of wider tyres (on standard rims) without the risk of tyre-to- spring contact. They are of two leaf,

Rear spring stiffener from the Ron Hopkinson handling kit.

lightweight design. Mk Parts sell similar offset springs.

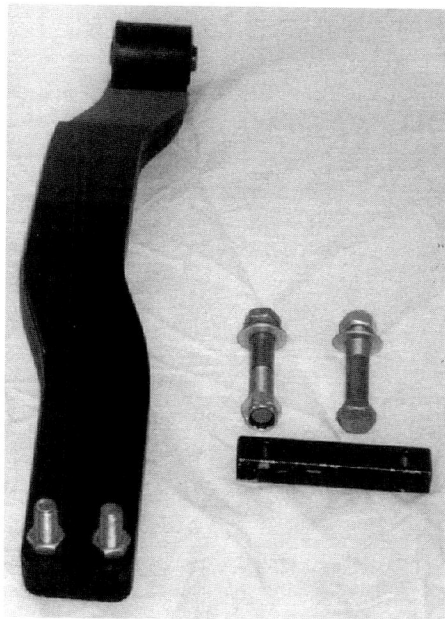

Offset rear quarter-elliptic spring. *(Courtesy MK Parts)*.

Before buying new springs have a think about the ride height of the car; if you lower it the centre of gravity is also lowered and this will reduce body roll and increase stability. As an alternative to buying reduced ride height springs, you can use lowering blocks. In effect, the block raises the axle in relation to the

Rear spring lowering blocks and U-bolts. *(Courtesy Victoria British Ltd)*.

spring, thereby achieving the same result as would a reduced ride height spring. The block is made of aluminium and is inserted between the axle and the spring. The car's intended use will decide the height reduction you choose (see option chart at the start of this chapter) but, whatever your choice, make sure it matches, or

exceeds, the ride height at the front of the car. Blocks come in the following sizes: 1in (25mm), 1.5in (38mm) and 2in (50mm). Extra long U-bolts for use with the blocks are normally included with the blocks but, if not, you should be able to get them from your local accessory shop. Kits are available from Motobuild and other Midget/Sprite specialists.

A better method of lowering rear ride height is achieved by raising the height of the front mounting plate for the spring (an extended mounting plate can be supplied by Peter May Engineering). The special plate raises the height of the front end of the spring which produces a reduced ride height of half-an-inch. However, it's possible to cut and weld a further extension and still retain the standard fixing point. Depending on how much the mounting plate is extended, it may be necessary to ensure adequate clearance by cutting and raising the height of the box section area the mounting plate sits in - this is only required for heights in excess of 1in (25.4mm), though. This method of lowering has the advantage of adding anti-squat characteristics to the rear suspension (when accelerating, the tyre is forced into the ground - by trying to dive under the pick-up point - causing greater weight transfer and reducing wheel spin and eliminating tramp). Following the theory through, you might expect that, by lowering the rear spring shackle, the same affect would be achieved. However, Peter May Engineering advises that this causes spring whip which is undesirable.

Although quarter-elliptic cars are generally not very well catered for in the UK, Tom Colby's Speedwell Engineering in the USA makes some suspension components for these cars, as do Faspec and Motobuild.

SHOCK ABSORBERS (DAMPERS)

The job of the shock absorbers is to stop spring oscillation (yo-yo effect). Its a good idea to change springs and dampers at the same time, and well before both are totally worn out. Incidentally, dampers have a

pronounced affect on braking distance if worn or too soft.

It's important to match spring and damper characteristics as, for instance, a stiff spring on a soft damper will make the spring work very hard and wear prematurely, and the soft damper will not control the spring effectively, whereas, a soft spring on a hard damper will have the damper trying to do some of the spring's work, again resulting in premature wear.

If you are sticking with lever arm dampers, once you've made your spring selections, you'll need to experiment with different internal valves in the dampers in order to obtain a balance you're happy with. It is, of course, much easier to get a satisfactory result if you are using externally adjustable telescopic dampers.

So, which is best, lever arm or telescopic? The answer depends on just how you use your car. Armstrong lever arm units may work well on the race track which is, after all, relatively smooth but they won't work as well on bumpy surfaces. It also has to be borne in mind that most Midget/Sprite races are quite short and this is significant because the Armstrong unit can fade, and thereby lose performance, in hard extended use. Generally speaking, telescopic damper conversions offer the best and most adjustable solution for most cars but, if originality or weight are primary considerations, you'll need to look at uprated lever arm units.

Adjustable lever arm shock absorbers

Now obsolete and scarce, you might nevertheless chance across the adjustable Armstrong lever arm shock absorber that was a BL Special Tuning part. They are very similar in appearance to the standard items, except for the large adjusting knob at the base. Old units are still reconditionable and worth buying if you find them. Fitting is straightforward as they are a direct replacement for the standard part.

Mini Mania produce a lever arm unit with an adjustable arm, this allows the

Rare adjustable Armstrong lever-arm shock absorber. (Courtesy Rick Neville).

alteration of camber angles.

TELESCOPIC SHOCK ABSORBER CONVERSIONS (FRONT SUSPENSION)

A conversion to telescopic shock absorbers is a good thing because not only do telescopic units have superior fade resistance, they're also available in a vast range of specifications. If you're a racer seeking the ultimate set-up for your car, which might vary from circuit to circuit, you'll find that the conversion kit manufacturers can advise on bump and rebound settings or supply purpose-built units to suit your own specification.

If, you just want to try different settings without changing shock absorbers,

Telescopic shock absorber conversion for front suspension available from Speedwell Engineering. (Courtesy Tom Colby, Speedwell Engineering).

you'll find it easy to do with modern, externally-adjustable, telescopic shock absorbers - this flexibility is rarely possible with lever arm units. One company with experience of the Midget/Sprite is AVO/Chassis Dynamics who retail the AVO range of on-car adjustable (bump and rebound) shock absorbers. If you're out to save weight, AVO can even sell you a unit in aluminium alloy. Remember that a weight saving on a suspension component is a saving in unsprung weight and is therefore more valuable than the weight saving of the component alone.

An important point on telescopic shock absorber conversions is that they may cause the brake hoses to rub against the road wheel when the wheel is on full lock. For this reason it's a good idea to use steel braided flexible brake pipes and to check regularly for damage.

Spridgebits/Speedwell kits
One method of converting to telescopic shock absorbers/dampers was the old Spridgebits kit which is no longer available. Despite its drawbacks, it did have the one great advantage of utilising on-car adjustable telescopic shock absorbers. In the USA an almost identical kit is still available from Tom Colby's company, Speedwell Engineering.

Should you obtain one of the old Spridgebits kits and the shock absorbers wear out, replacement shock absorbers are available from Spax Ltd, who supplied the originals, but quote the number of the old component, which is usually found at the base of the unit.

Frontline Spridget Ltd. kit
A superior alternative to the standard lever arm damper front suspension, the defunct Spridgebits telescopic kit and similar telescopic conversion kit from Speedwell, is the Frontline Spridget kit. This does away with the standard lever arm unit altogether and replaces it with a purpose-designed top link and telescopic shock absorber mounting. The second major component in the kit is the bottom link for

Frontline Spridget Ltd. telescopic shock absorber conversion kit showing all the major parts.

Frontline Spridget Ltd. suspension top arm in fitted position.

the telescopic shock absorber which ingeniously spreads the load across the wishbone, thereby reducing, if not

Frontline Spridget top arm mounting in fitted position.

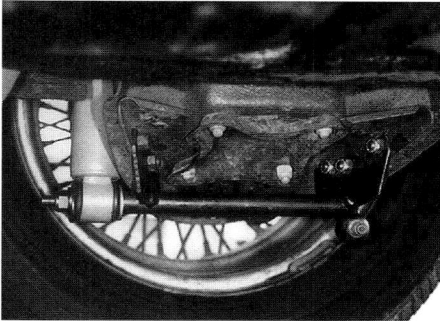

Frontline Spridget telescopic shock absorber kit fitted to a bottom wishbone.

Modified Frontline Spridget telescopic shock absorber kit fitted to Mk Parts' racing Frogeye. (Courtesy Mk Parts).

car adjustable in rebound only. At the time of going to press Frontline had just offered as an alternative an AVO shock that is on-car adjustable in bump and rebound and is specially valved to suit the Midget/Sprite; as a special option it's available as a lightweight version in aluminium alloy. However, in the unlikely event that the damping characteristics are not quite what's desired, AVO UK/Chassis Dynamics can rebuild the unit to a different specification. Returning to the kit, it comes with comprehensive fitting instructions.

If you are concerned that the kit might be a little bit on the heavy side, a back-to-back comparison with the old Spridgebits kit showed that, in fact, Frontline Spridget kit is fractionally lighter. However, the conversion is heavier than the standard Armstrong unit (a likely feature of any conversion to telescopic shock absorbers by virtue of the weight of the shock absorber itself). Of course, the real gains are in the improved handling the kit produces. One interesting feature of the kit is that the top link is designed with two degrees of negative camber built in which is just about right for road use. Any further increases in camber can, of course, be added by using Motobuild, or similar, negative camber bushes.

Overall, this is an excellent conver-

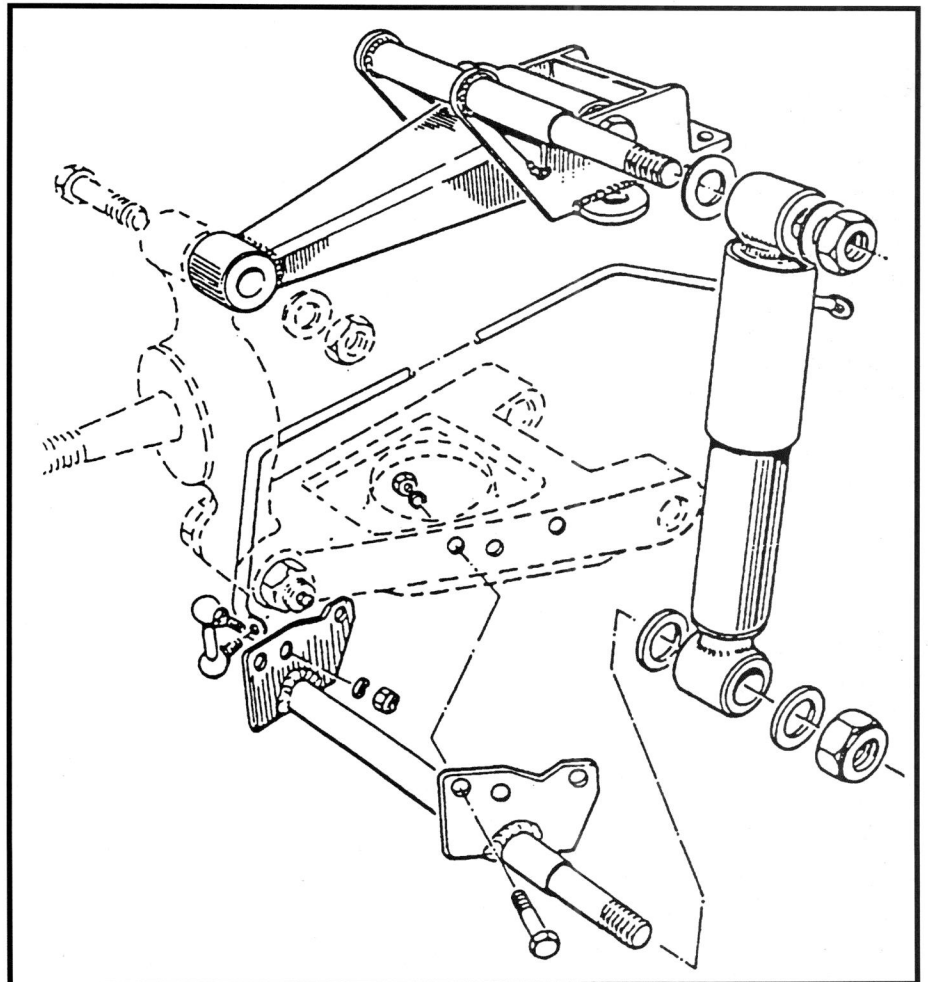

Frontline Spridget Ltd. conversion kit fitting diagram. (Courtesy Frontline Spridget).

eliminating, the wishbone breakages experienced with some kits.

Until recently, the shock absorber was purpose-made, designed for the kit and manufactured by Koni exclusively for Frontline. The Koni shock absorber is off-

A pair of the AVO adjustable shock absorbers. These are steel, but aluminium versions are available. Note large adjuster knobs.

Of course, if you want to be radical, you can build your own coil-over shock absorber front suspension like Eric Grundy's. (Courtesy Eric Grundy).

sion that is well engineered and worth fitting to any Midget/Sprite, especially with the AVO shock absorber option.

TELESCOPIC SHOCK ABSORBER CONVERSIONS (REAR SUSPENSION)

Spax kit

Spax manufactures a kit which replaces the Armstrong lever arm shock absorbers with adjustable telescopic units. Not only is the conversion neat and simple, it also provides dampers that are on-car adjustable. The telescopic conversion has been around for a number of years now and at various times has been favoured by the racing fraternity, although not so at the time of writing as the trend is back towards Armstrong lever arm units. It has also been suggested that the Spax telescopic conversion can introduce axle tramp - on this latter point it's a simple matter to fit anti-tramp bars. A variation on this conversion is to use the Spax kit but with AVO telescopic shock absorbers, which

Telescopic shock absorber conversion kit for quarter-elliptic cars from Spax.

are on-car adjustable in bump and rebound, whereas the Spax units (at the time of writing) are adjustable in rebound only.

The Spax units are on-car adjustable. Start at the hardest setting and work down to softer settings until you are content with the way your car handles.

It is worth noting that a Spax kit is available for quarter elliptic cars, too.

Frontline Spridget Ltd. kit

At the time of writing, Frontline Spridget Ltd. is producing a new telescopic

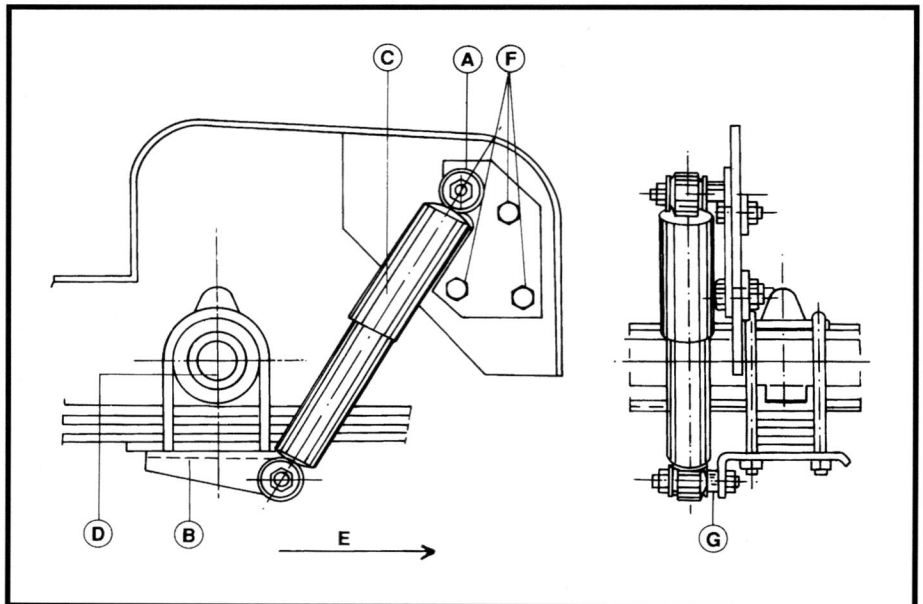

Fitting diagram for Spax telescopic rear shock absorbers conversion kit.
a) Upper bracket, b) Bottom mounting bracket against spring, c) Shock absorber, d) Rear axle, e) Front of vehicle, f) Top bracket securing bolts using existing holes, g) Bottom securing bolt for shock absorber (Courtesy Spax; redrawn by Sharon Monroe).

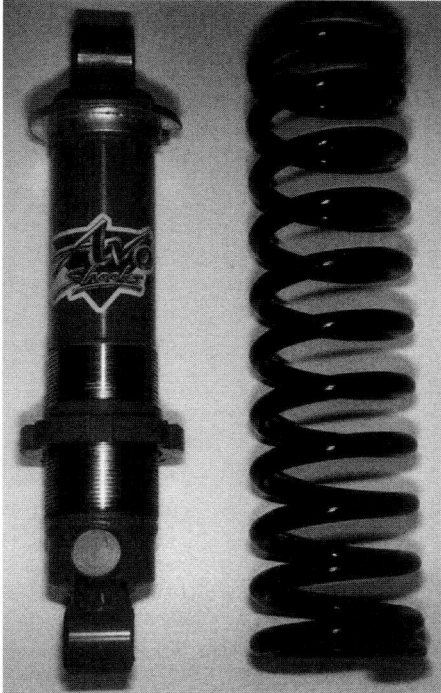

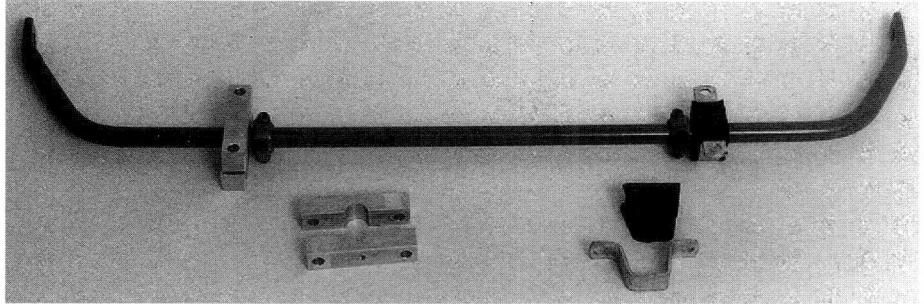

An anti-roll bar with both rubber and solid metal types of mounting block.

This is an AVO shock absorber with adjustable spring platform: ideal for a coil-over shock conversion.

Motobuild solid aluminium mounting block (fitted with hex-head rather than Allen screws).

conversion for the rear of Midgets/Sprites, including quarter-elliptic sprung cars. The kit is straightforward to fit with the shock absorber installed in a near vertical position, as opposed to the 45 degree angle of other kits. A further advantage is that this kit is lighter by over 10lb (5kg) than the Armstrong lever arm units.

This conversion is likely to be another landmark in the improvement of Midget/Sprite suspension.

ANTI-ROLL BARS

Front suspension

Having covered dampers, or, if you prefer, shock absorbers, it is important to look at one of the greatest areas for improvement in Midget/Sprite handling: that is fitting an uprated anti-roll bar and mountings. The biggest contribution to handling improvements will almost certainly come from the anti-roll bar.

There are currently at least five different thicknesses of aftermarket anti-roll bar available for the Midget/Sprite (see suspension options chart earlier in this chapter). The thicker the bar, the less the car will roll.

The standard anti-roll bar is mounted on rubber blocks bolted to the chassis legs. An alternative to rubber blocks are ones of made of solid aluminium which stiffen the ride even further. The aluminium blocks are direct replacements for the rubber blocks and the principle difference, apart from the obvious, is that they are secured by long Allen bolts rather than hexagon-headed bolts. In addition, they have a grease nipple, the head of which invariably breaks off!

Motobuild retail aluminium wishbone mounting plates, offering a weight advantage, for anti-roll bars. It's also possible to rose-joint the linkage from the mounting plate to the roll bar for added stiffness.

For many years the author's recommendation for the ultimate road set-up was an 11/16in (17mm) bar mounted in Motobuild solid mounting blocks, but he's now revised that recommendation in the light of further testing and the availability of stiffer bars. The Ron Hopkinson handling kit uses a 3/4in (19mm) bar in rubber mountings which works very well on a road car. To a large extent, this would appear to negate the use of softer bars. Of course, the Ron Hopkinson kit can be further improved by the use of Motobuild solid mounting blocks. On the other hand, you might want to try a 7/8in (22mm) bar in rubber blocks.

Most important of all, though, is to consider spring stiffness in conjunction with anti-roll bar thickness.

Rear suspension

As far as I know, only Mini Mania offers an anti-roll bar for the rear of the Midget/Sprite with kits for quarter- and half-elliptic sprung cars. The diameter of the bar is 5/8in (15.8mm). If you used this anti-roll bar kit, it would probably be quite easy to get bars made in different thicknesses to set up the car to your preference.

ANTI-TRAMP BARS AND PANHARD ROD (REAR SUSPENSION)

For cars putting out over 85bhp at the rear wheels and being used competitively, or simply very enthusiastically, the following are very worthwhile modifications.

When a car accelerates hard from standstill the wheels can twist the axle and wind-up the spring until the spring rebounds: the sequence then repeats itself. This phenomenon is known as 'axle-tramp' and can be seen when the driven wheels hop up and down during hard acceleration. Axle-tramp is not only detrimental to good acceleration, it can also break the differential casing. The total solution is to fit anti-tramp bars (available from Peter May Enginering) although, if the condition is not too bad, a Nylatron or Polyurethane bush kit can help alleviate it.

Anti-tramp bars

The anti-tramp bar is fitted between two brackets; the front bracket is bolted to the car's underbody below the front spring hanger plate, using existing holes, but slightly longer bolts. The rear bracket is fixed to the U-bolt clamp by passing the U-bolts through the bracket and re-tightening them. It is sometimes necessary to fettle front and rear brackets slightly to get a good fit. The principle area that needs attention will be the edges that sit over the spring, as they may impinge on the rivet of the spring clamp.

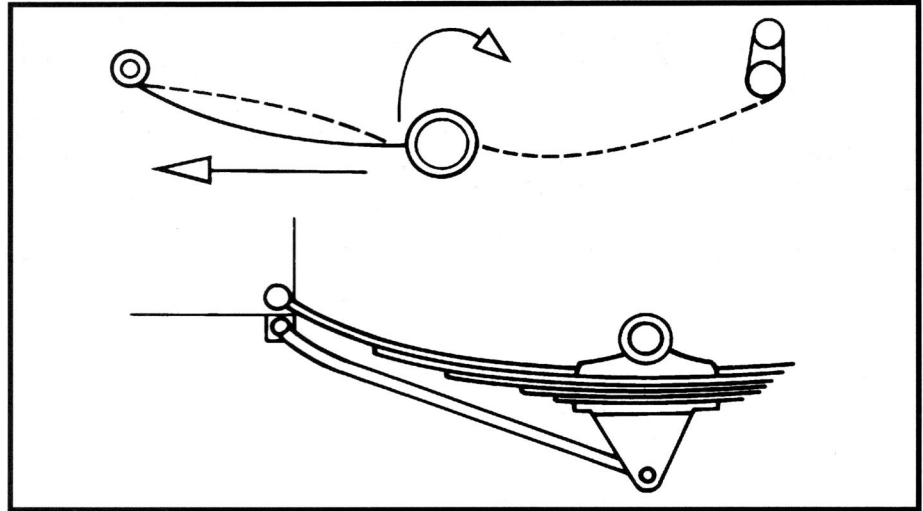

With standard suspension the rotational forces acting on the axle can twist or 'wind-up' the spring, causing axle tramp. This is prevented by the anti-tramp bar which restricts axle movement to rising and falling. *(Drawn by Dave Robinson/Sharon Munro).*

Anti-tramp bar combined with Spax telescopic shock absorber conversion by a special mounting.

The rear mounting for the anti-tramp bar.

If you use the anti-tramp bars with the Spax telescopic conversion, it may be necessary to grind the rear anti-tramp mounting bracket. This is done to give clearance between the bracket and the link arm. Alternatively, it's possible to weld anti-tramp pick-ups directly onto the shock absorber bracket, forming a single bracket to replace the two previously required. It's recommended that the centre hole in the base of the rear anti-tramp bracket be extended to the bar pick-ups by drilling through them. Check bracket to tyre clearance, and file or grind the centre bracket if necessary.

Before you fit the axle bracket, check to see how much thread is protruding through the existing damper link bracket, noting that you will need at least the thickness of the anti-tramp bracket before

Rear leaf spring, lowering block and anti-tramp bar.

really can be a do-it-yourself installation. The quality of the kit's engineering is extremely high.

Also available from Speedwell is a Panhard rod kit specifically for quarter-elliptic cars. This panhard rod kit is identical to that for the half-elliptic suspension cars, except for the fastening at the spring end of the rod.

Motobuild also makes a Panhard rod kit for quarter-elliptic cars, though the author has not yet had an opportunity to see one.

a successful fit can be achieved. Although the 1500cc-engined cars have more leaves than those on the 1275cc-engined cars, only one length off U-bolt is generally offered with this kit. Even so, lengths can vary from supplier to supplier and, with an anti-tramp fitting, a minimum length of 5.9in (15cm) is required and that 5.7in (14.5cm) lengths will be difficult, if not impossible, to fit. This is particularly the case if you are using Nylatron bushes which will not compress like rubber bushes. A longer, off-the-shelf, alternative to the Midget/Sprite U-bolt is the MGB U-bolt which is ideally suited. It is thought that the MGC U-bolt is even longer. Finally, an added bonus of using anti-tramp bars is that they stiffen-up the rear of the car, increasing resistance to roll.

Panhard rod

On the Midget/S prite rear suspension it is the leaf springs which support the axle and provide control of sideways movement of the body in relation to the wheels. This is not an ideal arrangement because the control is limited and, once wider wheels and tyres are fitted, the arrangement's shortcomings become more apparent.

The solution is to fit a Panhard rod kit. The rod is attached to the body at one side of the car and to the axle on the other side of the car at the other end. Each end of the rod finishes with a rubber bush or rod end spherical joint. The brackets that mount the rod are such that the rod is held parallel to the axle when fitted correctly.

Peter May Engineering, Motobuild and Tom Colby's Speedwell Engineering

(in the USA) can supply a Panhard rod kit for the Midget/Sprite. Also, note that Speedwell parts are generally available from MK Parts in Germany. Some kits are specific to Armstrong shock absorbers or telescopic conversions, so specify which you are using when you place your order.

If you are using tyres wider than 165, the fitting of a Panhard rod kit is recommended. The Speedwell kit uses a tubular rod with rod end spherical joints. Note that if you are using rod end spherical joints on a road car, you should fit protective rubber gaiters to them, otherwise everyday road dirt grime will dramatically shorten the life of the joints.

The Speedwell Panhard rod comes with a very comprehensive set of instructions. Another important point is that no welding is required in fitting the bar, so it

SUSPENSION GEOMETRY

An accurate total alignment check is essential if you want to set up your road or race car correctly though, unfortunately, you cannot usually get the required high level of measurement quality from your local tyre specialist. You need to find a company equipped with sophisticated equipment which works on the basis of light beams and which can accurately measure the alignment, in all planes, of every wheel. In the UK, Pro-Align of Northampton are able to offer such a service (and the company's service manager, Jim Irwin, has raced a Midget in past years).

Before taking the car along for the

Speedwell Engineering Panhard rod kit (available for quarter- and half-elliptic-sprung cars) fitted to MK Parts' racing Frogeye. (Courtesy MK Parts).

Pro-Align wheel alignment equipment.

geometry check, it is recommended that tyre pressures are correctly set. Also, ensure wheel bearings and suspension are in good order, since any problems here negate the accuracy of the measurements and any benefits from the set-up following alignment.

Once your car is in the workshop it is put on a special ramp and (in the case of the Pro-Align check) infrared computerised measuring heads fitted to the wheels. The wheels are first checked for run-out (buckled or bent). The next series of checks produces a computerised print-out of toe, castor, king pin inclination and wheel set-back for the front wheels and toe, thrust angle and camber for the rear wheels.

The author took his road legal, modified Sprite along to Pro-Align for a full alignment check and, after the readings had been taken, discussed the results with Jim Irwin. The (unexpected) bad news was that the equipment had detected that the rear axle was slightly bent (though within acceptable limits). The tracking was out, but this was easily corrected and, most interestingly, the camber and castor settings were out each on opposite sides of the car and, because of this, self-cancelling to the point where the car should drive

straight (it does).

The biggest thing the author learnt from his visit to Pro-Align was that fitting a collection of uprated new parts will not necessarily give the required result. For instance, the suspension on his car, including a Frontline conversion, should have produced a certain amount of negative camber in conjunction with the lowered springs, yet, because the car was no longer factory perfect, it was found that the car would benefit from a little more negative camber on both sides (slightly more on one side than the other). Having a smaller amount of negative camber than was thought, also explained why the Frontline kit made such a positive difference at high cornering speeds. It's likely that, before fitting the Frontline kit (which introduces negative camber in the suspension) the car must have had a non-standard and excessive amount of positive camber. Of course, no two cars are the same, a fact which further underlines the value of a really good geometry check.

Tracking (toe-in & toe-out)
The standard tracking setting for the Midget/Sprite is an 1/8in (3.17mm) of toe-in. If you have fitted wider wheels, this

needs to be reduced for optimum handling and it's suggested you reduce toe-in by 1/32in (0.79mm) at a time and see how the car feels. For a race car on wide tyres, the optimum setting is likely to be between 1/32in and zero: however, as with the road car, it's ultimately down to driver preference.

Camber
Lowering the front suspension changes the wheel camber - camber being the angle at which the wheel leans inward or outward in relation to the road surface. The standard Midget/Sprite is set with approximately three-quarters of a degree of positive camber. Since neutral to negative camber will improve handling, new camber angle settings can be achieved by a variety of means.

Under hard cornering conditions the outside, loaded, wheel should be as near vertical (in relation to the road surface) as is possible in order to achieve maximum grip; this is the objective of camber adjustment. Before setting the camber, note that if you have lowered the car the camber will have been reduced from positive to negative. Each inch (25.4mm), approximately, of reduced height can equate to no more than 1 degree of camber change (moving towards negative from positive).

However, it's possible to have too much of a good thing, and too much negative camber can cause a serious reduction in straight line braking performance, introduce a tendency to high speed wander, create excessive bump steer, reduce cornering efficiency and introduce a great deal of oversteer. So, how much is too much? Well, for a road car on wider than standard wheels, a total of 2 degrees negative is recommended. It's a good idea to have a high quality alignment check before making adjustments, and then set camber according to the results obtained and your preferred driving style. An expert wheel alignment check may reveal that the car would benefit from camber adjustment to compensate for errors elsewhere in the overall wheel alignment set up (eg:

different castor settings on each side of the car). It's hard to say how castor and camber - which, in theory, are set at the build of the car - can vary but the cars are old now and, over the years, road bumps and bashes can alter chassis and wheel alignment.

One method of changing camber is to replace the rubber trunnion bushes with special bronze bushes which allow camber adjustment as the centre hole is eccentric. Amongst others, Mini Mania market an appropriate kit but theirs has the advantage of locking the chosen settings with pinch-bolts.

Eccentric brass trunnion bushes give adjustable negative camber: from Motobuild.

Aldon Automotive makes cast alloy trunnions that produce negative camber and Gillspeed, in Australia, sells a negative camber trunnion, too.

Another method of camber adjustment is to use the Mini Mania competition lever arm shock absorber which allows for camber adjustment.

Mini Mania modified lever-arm shock absorber with rod end fastening. (Courtesy Mini Mania).

A further consideration with regard to camber is body roll, since less roll means less camber change at the wheels during cornering. For example, a road car on soft suspension settings may need more negative camber to produce an upright wheel in a tight bend than a stiffly set race car that has much less body roll. With this

Speedwell Engineering lightweight alloy rose-jointed radius arms fitted to MK Parts' racing Frogeye. Note special three-leaf spring and finned alloy brake drum. (Courtesy MK Parts).

in mind, camber may need adjustment each time changes are made to ride height, body weight and roll stiffness (springs and anti-roll bars).

A final point on camber. A road car benefits from a slight bias in nearside (wheel nearest kerb) wheel camber to compensate for the curvature of road surfaces, whereas a race car would be set to be neutral.

NEW WISHBONES (FRONT SUSPENSION)

If your car's front suspension wishbones need renewal, always ask for replacement components with anti-roll bar holes pre-drilled, whether your purchase is new or reconditioned wishbones. The reason for this is that, even if your car doesn't have an anti-roll bar currently, when you do get around to purchasing and fitting one, you'll not need to do any drilling, saving both time and possible trouble. Ron Hopkinson offer a service where new or reconditioned wishbones can have a second grease nipple brazed in, a modification which improves the lubrication of the fulcrum pin.

RADIUS ARMS (REAR SUSPENSION)

The standard rubber-bushed radius arms (trailing arms) can be replaced by aluminium arms using rod end spherical joints. The benefit of using these is a weight reduction as well as firmer location provided by the rose-joints. The Speedwell radius arms are fully adjustable at each end, and of fine quality construction. In the UK, Motobuild also manufactures rose-jointed radius arms.

SHOCK ABSORBER LINK (REAR SUSPENSION)

Mini Mania appears to have a unique product in a tubular, aluminium heim-jointed damper link for quarter-elliptic cars. Note that the principle difference between the US heim-joint and British rose-joint is that the former has grease nipples. At the time of going to press no similar link existed for half-elliptic cars, though it been suggested to Don Racine at Mini Mania that he should make some!

Standard anti-roll bar link.

Rose-jointed, adjustable anti-roll bar link on Simon Page's turbocharged Frogeye Sprite.

UPRATED SUSPENSION BUSHES

When rebuilding your car's suspension you may wish to consider replacing all the standard (and very compliant) rubber bushes with bushes made in Nylatron, polyurethane or bronze. These types of bushes will produce a much stiffer and tauter feel to the suspension. However, the downside is that you might find the reduced compliance increases noise, vibration and harshness (NVH) to an unacceptable level for day-to-day use. Your circumstances and the other modifications you have made to the suspension will dictate whether you take this option or not.

It's also possible to replace the anti-roll bar link with a metal rose-jointed items.

Nylatron bushes

Before purchasing replacement rubber bushes - including the ones between the springs and plates and for the all the suspension pick-ups on your car - consider using Nylatron bushes. This type of bush offers greater resistance to movement in the suspension and will give a tauter feel. The Nylatron bushes will, unfortunately, also give a harder ride as they transmit more shock than a rubber bush. The change to Nylatron from rubber is a competition rather than road modification, although you may feel the negative aspects are small enough to live with on your road car. The bushes are available from Motobuild Ltd. Fitting should be

Nylatron suspension bushes.

Nylatron bushes for rear springs.

straightforward but the bushes may need a little fettling with a file or rasp to get them to fit - do not force them. To further aid fitting, lubricate the bushes with copper-based grease.

Polyurethane bushes

An alternative to Nylatron or rubber bushes in the suspension are polyurethane bushes, which are much less compliant than rubber but less brittle than Nylatron. They will last an awful lot longer than rubber because they are resistant to oil, grease and road salt. You can buy them from Faspec in the USA, Gillspeed in Australia and Frontline Spridget Ltd in the UK. Kits are available for wishbones, trunnions and rear suspension.

Polyurethane bush kits are available from Faspec in the USA - these are for the rear of quarter-elliptic cars.

COMPETITION STUB AXLE (FRONT SUSPENSION)

Although examples are rare, apparently, the standard stub axle can break. Mini Mania has a competition stub axle manufactured in high grade steel and with generous radii at critical areas. These competition stub axles are pressed into the stub axle carrier and then welded in. They are claimed to be unbreakable

ALLOY WHEEL HUBS

It's possible on steel-wheeled cars to replace the standard hubs with lighter alloy

Competition stub axle from Mini Mania which clearly shows the welding on the back of the spindle. *(Courtesy Mini Mania).*

ones; such hubs are available from Merlin Motorsport in the UK and Mini Mania in the USA.

STEERING MODIFICATION OPTIONS

Most people don't do much with the Midget/Sprite steering system as, even in standard form, it works very well. However, a real problem with the steering is the driver's lack of arm/elbow room. As a solution a lot of the racers use a dished steering wheel fitted the 'wrong' way around, but a better solution can be found in shortening the steering column.

In general, all that needed to make the most of the steering system is to make sure that all components are in good order - including the rack mountings.

The steering can be 'quickened' by fitting a smaller steering wheel, but this will also make the steering heavier.

STEERING RACK

Late model Midgets (1972 on) used a different rack to that of the early cars. Of Triumph Herald/Spitfire origin, it is 'slower' than the early racks and requires more turns lock-to-lock. There is, however, only a negligible difference in steering response and a smaller diameter steering wheel would do more to sharpen up the general feel of the car, than a swop of racks. If you do decide to swop racks, you'll also need to change the rack mountings, tie rods and steering arms.

SHORTENING THE STEERING COLUMN

How much of a reduction can be achieved is not so much limited by the inner column as by the steering boss. Also, the column and boss must have very near equal amounts of material removed. The reason for this is that the internal splines of the boss are near the rear face of the boss, so it's only really practicable to remove an 1in (25.4mm) or so, but the safe amount will vary from boss to boss. You certainly shouldn't remove more than half the depth of the boss' original splines.

When the column comes back duly shortened, it can be fitted it into the car as per the workshop manual instructions. Once the column is back in place, the boss iss turned down and the clearance between the indicator shroud and the boss checked. If the space is insufficient, it is a simple matter to cut and file the shroud to fit snugly against the boss.

CAUTION: If you do decide to shorten the steering column, be absolutely sure that the engineering company you choose are expert welders, as weld failure will result in loss of steering.

When you refit your shortened steering column you may need to modify the horn push connection that is fitted to the indicator switch assembly, depending on how much you have had the column shortened. You'll see that there is a short strip of brass with a miniature round button-like copper contact at the end, which is designed to make contact with the brass ring seated at the top of the column. With a shortened column you may well find that the contact is now too far up the column relative to the brass ring. Use a

This is the end of the steering column to shorten.

Shortened steering column fitted in the car and steering wheel boss machined down to fit against face of indicator shroud.

small file to file the reverse edge of the copper contact until it can be pulled free of the brass strip. Next, drill a hole lower down the brass strip and refit the copper contact. The reverse side of the copper

A steering wheel which is flat, rather than dished, gives the driver more elbow room.

contact can be lightly hammered to rivet over the contact to secure it.

If you plan to use a super short steering column you'll need to cut the scuttle panel back as required.

Trimming the indicator shroud has already been mentioned, but it will also be necessary to modify the indicator stalk itself to keep it within finger-tip reach. This can be done by removing the stalk and flattening out the kink till it is straight.

Other column modifications

A useful modification to the steering column installation on the Midget/Sprite was contributed by Nik Handford of Omni Autos. Junk the outer steering column sleeve, and run the column in a nylon bush (still using the original mounting bracket) the result is elimination of play and a modest weight saving. Nik Handford and Simon Page carried out this conversion on Simon's racing car with few problems.

Probably the ultimate steering column is available from Mini Mania in the USA. The stock column is modified to incorporate a U-joint and is supported by a spherical bearing. The column is also fitted with a quick release hub for the steering wheel.

STEERING WHEEL

There is a vast selection of steering wheels available for the Midget/Sprite, including wooden-rimmed wheels for that authentic 'sixties look. The wood-rimmed wheel is a Moto Lita item of 13in (330mm) diameter in a flat style which gives an extra 2in (50mm) of driver elbow room over that of a standard car. However, be aware that in racing circles, although legal, wood rim wheels are frowned upon.

The smaller the diameter of the wheel you fit, the more direct (quicker) the steering becomes. The downside is that the steering will also become heavier as the wheel becomes smaller: the answer is a sensible compromise that suits you and your car.

Chapter 10
Wheels & Tyres

WHEELS

Wider wheels

Both wheels and tyres come in different widths and diameters, but in all cases the original standard diameter of the Midget/ Sprite roadwheel is thirteen inches. Rim width is normally 4 inches (101.6mm) or 4.5 inches (114.3mm). Normal tyre width is 145.

Your car will have pressed steel (maybe Rostyle) or wire wheels, assuming you haven't already changed them for non-standard items. If you're looking to use any tyre width greater than 155, then a wider wheel should be considered. Unfortunately, fitting wider wheels to the Midget/Sprite is difficult, as it will almost certainly require wheelarch modification: read the bodywork chapter for further details.

For the replacement of pressed steel (including Rostyle) wheels there is a wide selection of aftermarket products in both alloy and steel; however, before you make your final choice, consider the wheel's offset. Aftermarket wheels usually have the offset to the outside edge, so appropriate

Alloy wheel styled upon the famous Minilite which was originally made from magnesium.

wheelarch flaring can accommodate the extra width, but do check this before purchase.

A fine wheel is the alloy Minilite replica, retailed by Motobuild and available in 5.0 or 5.5 inch rim width. As well as the bolt-on Minilite replica, Motobuild can supply a Minilite replica that can be used with knock-off hubs. As well as providing an instant conversion from wire-spoked to alloy wheels, this particular

wheel has the advantage of being adjustable for offset and inset via a spacer between the main wheel body and hub insert. If you want the real thing, genuine Minilite wheels are available once again.

Revolution wheels in 5.5 or 6 inch rim sizes are available. Peter May Engineering Ltd. offers a choice of 6 inch rim, 8-spoke steel wheel or a 6 inch rim KN diamond alloy wheel.

Whichever alloy wheel you use, note

Note the spinner on this genuine, if tired, Minilite is lockwired to the wheel to prevent it working loose.

that they usually require special wheel nuts. Also, in view of the cost of alloy wheels, it's recommended that you purchase a set of locking wheel nuts to protect them.

Wide wheels and tracking (toe-in/out)

Note that if you increase the wheel width - be it on a road or race car - you will need to reduce the toe-in. For further details see the suspension & steering chapter.

Wheel weight

Usually, not enough thought is given to the weight of the wheels and why wheel weight is important.

Reducing the wheel weight reduces what is known as the 'unsprung' weight of the car which, in turn, improves general handling and response.

Of course, wheel weight is also part of the mass of the car and, therefore, has an affect on acceleration. However, there is a further, more subtle consideration: as each wheel is a revolving part, its speed of rotation has to be accelerated (and decelerated) too. There is an obvious analogy with the importance of reducing the weight of the rotational parts of an engine - especially the flywheel.

Taking all these factors into account, you'll see that reduced wheel (and, for that matter, tyre) weight has a considerable affect on the power to weight ratio of your car.

Wire wheel conversion

To convert the car from steel or alloy to wire-spoked wheels, you need to change the back axle and front hubs or purchase a conversion kit from an MG specialist. The principle point is that you cannot just swop the hubs of a bolt-on wheel car with wire wheel hubs as, amongst other things, the halfshafts are different. An alternative which does not involve hub and halfshaft swapping, is to fit bolt-on wire wheels which are available from Motor Wheel Service (MWS).

This is a Minilite replica that has a steel insert to enable it to be used with knock-off hubs, available from Motobuild.

MWS 5.5 inch rim-width wire wheel fitted with Goodyear 175/70 NCT tyre.

The drawback with wire wheels is that you normally have to retain the standard, relatively narrow, rim width. However, it's possible to have wide rim wire wheels made. The width goes from 4 inches (101mm) to 5.0 or 5.5 inches (127 or 140mm) which is quite adequate for road use. The wider wheel has an outside offset, so will require wheelarch flaring. One company that can build these wheels to special order - in either painted silver or chrome finish - is MWS. MWS has worldwide distribution and its products can be purchased from Victoria British (Long Motor Corp.) and Moss Motors in the USA. Motobuild retails wide rim wire wheels, too.

Wire wheel owners should consider the replacement of the standard two-eared

Chrome wire wheel 3-eared knock-off spinner. (Courtesy Victoria British Ltd).

Lightweight spinner, produced by Motobuild.

knock-off wheel spinners by the MWS three-eared type; they do exactly the same job, but look smarter. If you're historic racing, you can at least save a bit of weight by using either hexagonal or two- or three-eared lightened spinners which can be supplied by Motobuild.

A final point on wheel choice relevant to the square-arched Midget/Sprite is that it can be impossible to fit certain wide wheel/tyre combinations without flaring the arch, so be warned. The practical tyre size limit is 165/70 on standard rims and

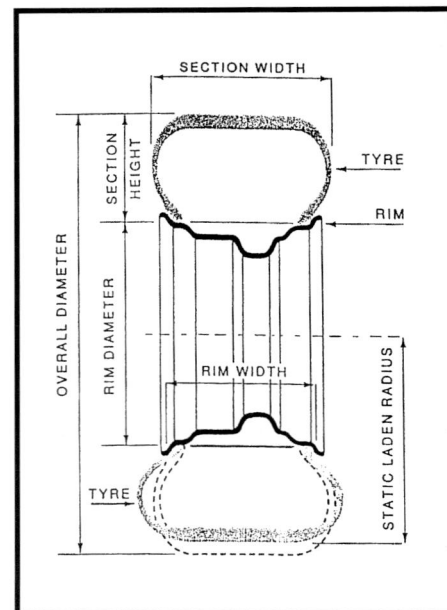

Anatomy of the wheel and tyre.

less if the aftermarket wheel has significantly more offset than the standard wheel.

TYRES

The subject of tyres is a big topic, but there are some things always worth remembering: do always use the same tyre widths front and rear as unequal widths give some undesirable, if not dangerous, handling characteristics such as sudden understeer/oversteer; don't try and fit low profile tyres to wire spoked wheels - they are not designed for them; do make sure the tyres you buy have an adequate speed rating for the speeds you have modified your car to attain as failure to do so can invalidate your insurance; do check that through full spring travel and wheel lock the car has no tyre fouling on bodywork, flexible brake line or suspension components; don't mix crossply (bias) and radial tyres on the same axle and, preferably, not at all

Tyre selection

Goodyear and Michelin gave advice regarding tyre selection as well as some useful tyre tips. Goodyear provided a quick reference chart for rim and tyre width suitability which is reproduced nearby. Michelin provided a diagram which 'decodes' the reference data contained on a tyre and were also kind enough to supply wheel revolution per mile figures for a selection of tyre sizes. The revs-per-mile figures are needed when calculating speed potential for various gear and differential ratio combinations. Incidentally, it's not important whether or not the tyres are worn in the tread, because tyres tend to 'grow' slightly with use due to centrifugal force and any differences between new and old tyres is negligible.

Should you choose to ignore the advice in the accompanying Goodyear tyre/rim chart, your tyre fitter will experience severe difficulty fitting an oversize tyre onto the rim, but it can be done. However, while, for instance, it's possible

Optimum tyre & rim combinations				
Tyre size	Minimum rim width	Wheel revolutions per mile	Optimum rim width	Maximum rim width
155R-13	4.00ins	-	4.50ins or 5.00ins	5.50ins
155/70R-13	4.00ins	958	4.50ins or 5.00ins	5.50ins
165/70R-13	4.50ins	931	4.50ins or 5.00ins	6.00ins
175/70R-13	5.00ins	909	5.50ins or 6.00ins	6.00ins
185/70R-13	5.00ins	882	5.50ins or 6.00ins	6.50ins
185/60R-13	5.00ins	955	5.50ins or 6.00ins	6.50ins

to fit 175 width tyres onto standard 4 inch rims this combination produces an unpleasant, sloppy, feel to the car when cornering hard. Although this may sound like a modest problem, it can feel alarming when driving at speed through a sequence of bends. 5.5 inch rims are much, much better for this size of tyre.

Like wheels, tyre choice will, ultimately, be down to personal preference and availability. In case it's of interest, the author's current tyre choice is the excellent Goodyear range of NCT tyres. He's used the original NCT in 175/70-13 size on 5.5 inch rim wire wheels with very satisfying results, but now runs his car on NCT3s in the same size and on the same wheels (the similar NC2 is better for town traffic/wet driving than the 3).

Also recommended are the Goodyear Aqua and GT ranges which are available in 155/70, 165/70 and 185/70 widths. These tyres have a speed rating of 118mph which should be sufficient for all but the very fastest of tuned Midgets and Sprites. As its name suggests, the Aqua is designed for wet weather: it has a deep central channel is of uni-directional design.

Remember to have the wheels balanced when you have new tyres fitted.

The accompanying chart is a guide to optimum wheel/tyre combination sizes.

Tubeless tyres & wire wheels

It's not generally recommended to use tubeless tyres with inner tubes. However, it's pretty much a necessity with wire spoked wheels (the author has not experienced any problems in doing this with his car).

When you go to your local tyre specialist, you may discover that they have never fitted tyres to wire wheels and don't have a selection of tyres specifically designed for use with inner tubes. In this case, you can advise the fitter that the inner tube should be coated with plenty of French chalk and inflated to its natural shape before insertion into the wheel and tyre. This will ensure the tube is correctly positioned and in full contact with the tyre before final inflation. Also, it's a good idea to fit new rim tape each time you new tyres are fitted (tank tape, available from many motorsport retail outlets, can be used).

You may have heard (especially in the USA) of wire wheels that have had the spoke sockets sealed with mastic to enable tubeless tyres to be used without an inner tube. However, only specially designed rims can be mastic sealed and, unfortunately, mastic cannot be used on the Midget/Sprite wheel due to the shape of the wheel profile where the tyre bead sits. To try and seal the Midget/Sprite rim is not only illegal (in the UK, at least) but dangerous.

A final point about wire wheels is that Michelin advises that, in no circumstances, should inner tubes be fitted with tyres of less than 65 per cent aspect ratio. The reason for this is that problems of air trapping and chafing can lead to premature failure of the tube, or even the tyre.

TYRE PRESSURES

Tyre pressures should always be checked cold.

Dunlop racing tyre compounds	

DRY		WET	
Compound	Usage	Compound	Usage
204 (hard)	Vintage		
484	Vintage	484	Vintage
764	Metro Challenge		
770	Medium hard/saloon	430	Saloon/Rally (very soft Rally)
		452	Sportscar
622	Rally		
476	Soft Endurance/ Saloon/Rally		
436	Soft Rally		
430 (soft)	Very soft Rally (hard wet)		
		186	F3
922	Qualifier		

Generally speaking, an increase in tyre pressure will improve the handling of the car but have a detrimental affect on ride quality and tyre life. Ten per cent over standard pressures is usually appropriate for fast road driving though, in some instances, better performance may be achieved by reducing tyre pressures by 2 to 3psi (13.79kpa to 20.68kpa). If you use non-standard tyre pressures, be guided by the manufacturer's recommendations and local legislation. Over-inflation causes excessive tension in the casing cords, which makes the tyre more vulnerable to impact damage and increase wear. Under-inflation allows excessive flexing and rapid overheating, leading eventually to casing break-up and failure; it also increase tyre wear.

Note that the effect of tyre pressure on the performance of the vehicle is not always the same for handling as it is for grip. For example, high tyre pressure may well improve handling due to greater stability of the tyre walls, but will reduce grip as the tyre contact patch, or 'footprint,' will be smaller. You need to decide whether your car needs better handling or more grip to go faster. For race cars, which have better handling than a road cars, an increase in grip usually yields better results than an improvement in handling. For road cars, it's likely that the reverse will apply, but only testing will provide the answers.

For tyres used in motorsport there is, unfortunately, no golden rule regarding optimum pressure. It is recommended that experimentation starts at 20psi/138kpa (cold) and ends at a maximum pressure of 32psi/220.6kpa (hot).

Generally speaking, to reduce understeer you should increase front tyre pressures while reducing rear tyre pressures. To reduce oversteer, the opposite applies. Of course, many other factors can affect these handling characteristics, too.

Tyres for racing

Tyres for circuit racing will fall into four groups: treaded race compound tyres, slick race tyres, wet race tyres and rally special stage tyres.

Where most road tyres are of radial construction, circuit racing tyres that are slick or wet are just as likely to be of crossply (biasply) as radial construction. The principle characteristics of the two tyre types are as follows. A crossply tyre has fast warm-up, due to tread squirm and excellent point-in. Conversely, radials give better traction, slower warm-up and poorer turn-in but, for long races, extended stability of performance. Obviously, the type of racing you do will dictate your choice but, as a guide, sprint and circuit cars are likely to be best served by crossply construction tyres, whereas cars competing on long tarmac special stages or in an endurance circuit race would benefit from the use of radials.

Just like in Formula One, racing slick race tyres come in a variety of compounds. A hard compound is long-lasting but will produce less grip than a soft compound. Wet race tyres are 'cut' slicks with increased rubber thickness. Race tyres are of a lighter construction than road tyres and, generally, only have two plies which makes them lighter and therefore easier to accelerate. Each manufacturer has its own compound rating system and to give an indication of what is available the accompanying chart shows the most common Dunlop compounds.

Treaded race compound tyres are soft compound, road legal race tyres of radial construction. This is the type used in one-make championships.

Rally special stage tyres will either be for forest or tarmac use. The selection of

Producing a wet racing tyre. Once a pattern has been spray painted onto the slick tyre through a stencil, the tread pattern is cut out by hand.

Wet and slick side by side (the slick appears smaller as it's deflated).

about rotating the tyres is that it can sometimes mean the tyre is running aga   ommended direction of rota   arked on the sidewall of the tyre). Only trials with your own tyres and car will decide if these latter points are a problem; in particular note that running against rotation is against the manufacturer's advice, though they are well aware that many teams do this to gain maximum usage from their tyres.

You may have heard of the phrase 'scrubbing' in relation to tyres used in motorsport. This describes the process whereby new tyres are run on the car for two or three laps of a circuit at a moderate speed to bring them up to temperature, and then allowed to cool. This heat cycle causes the tread compound to harden slightly, which makes the performance

more consistent over a full race distance, although the absolute best lap time will be slightly slower.

Sometimes people confuse scrubbing with buffing and use the two phrases to mean the same thing, which is not correct. Buffing is when a road tyre is shaved down to, say, 0.15in (4mm) tread gauge from brand new. The reason for this is that a full depth treaded tyre can suffer from overheating on the outside shoulders when raced. Buffing will also increase tread block stability. Generally, a tyre prepared in such a way will last longer and perform better on the race track than a new tyre. However, while using buffed tyres might be excellent in the dry, it's no good in the wet and you'll therefore need a separate set of full-treaded tyres for wet racing.

forest tyres is more a question of tread pattern and width than anything else.

Once you have your racing tyres, you can use the tyres in the same position all season or rotate the tyres around the car. The former will identify which tyre wears quickest, suggesting either the use of a harder compound on that corner or, conversely, softer compounds on the other three. Rotating tyres around the car after every race allows for even tyre wear throughout the racing season. However, because of the large amounts of camber used on the Midget/Sprite, the wear is uneven so, although rotation around the car will improve wear, it may be slightly detrimental to performance as initially, in each new position, the tyre contact patch will be reduced in size. The other point

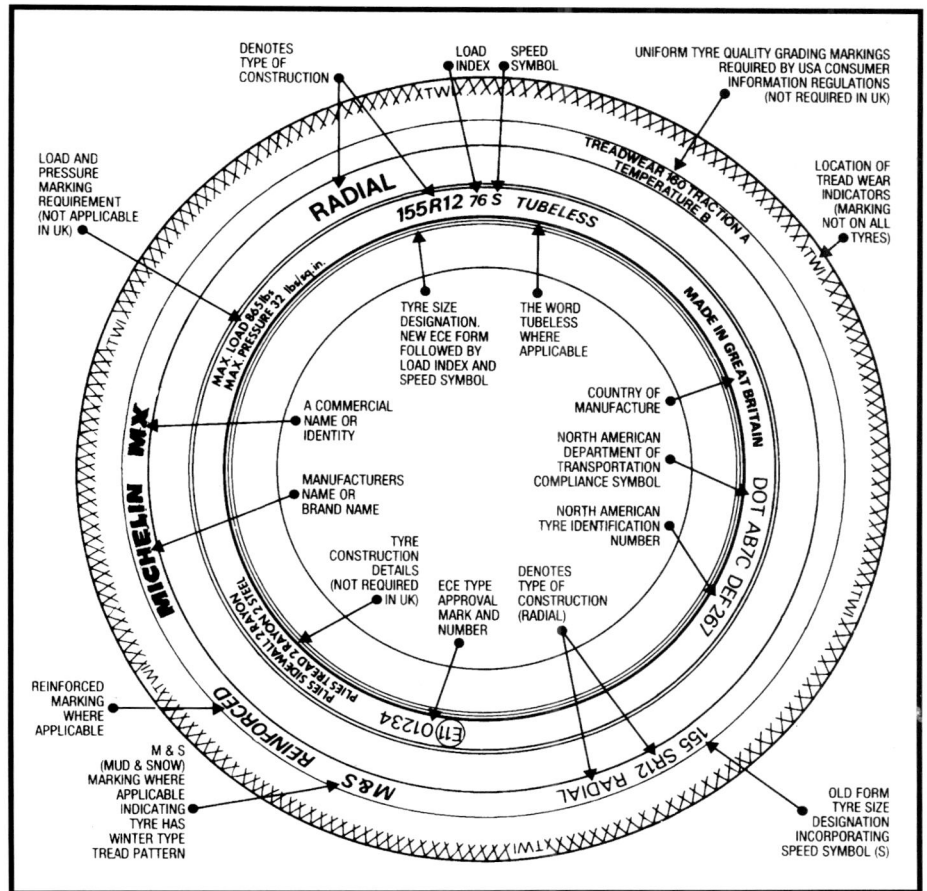

'The writing on the wall' - See text for explanation of symbols and codes. (Courtesy Michelin).

Tyre markings

'The writing on the wall' illustration shows all the markings which may appear on a tyre sidewall and covers both European and North American standards. The markings likely to be of most interest are those which indicate size, speed and load rating. The form of such markings has changed in recent years. The old markings on a radial-ply tyre of standard, nominally 80 per cent aspect ratio, take the form "165 SR 13," where "165" is the tyre's nominal section width in millimetres, "S" indicates the speed rating (more of that later), "R" indicates that the tyre is radial-ply and "13" is the nominal rim diameter in inches. A tyre with a different aspect ratio (ratio of nominal section height to section width) under this system would be marked, for instance, "165/70 SR 13" where "70" indicates a 70 per cent aspect ratio (nominal section height 70 per cent of nominal section width rather than the 'standard' 80 per cent). The new marking takes the form "165/80 R 82T MXT80" where "165," "R" and "13" have the same meaning as before. "80" is the aspect status. "T" is the speed symbol. The figure "82" is the load index (of little interest to us) and "MX" is the tread pattern.

The speed symbol on the tyre indicates the maximum speed at which a tyre can carry the load corresponding to its load index under specified conditions which are decoded as follows:

"Q" symbol: maximum 160km/hour (100mph)

"R" symbol: maximum 170km/hour (106mph)

"S" symbol: maximum 180km/hour (113mph)

"T" symbol: maximum 190km/hour (118mph)

"H" symbol: maximum 210km/hour (130mph)

"V" symbol: maximum 240km/hour (150mph)

Chapter 11
Braking System

MODIFICATION OPTIONS

Given that all the components of the system are in good order and properly adjusted, the brakes of the Midget/Sprite are good, even in standard form. The reason the standard brakes are effective is that the car, as befits a sports car, is light in relation to the size of its brakes (less so the 1500 Midget). The braking system is in sound condition when it meets the following minimum criteria. Front discs need to be free of corrosion, scoring and be of adequate thickness. Rear drum friction surfaces need to be corrosion-free, round and of adequate thickness. Drum brakes must be properly adjusted. Both disc pad and brake shoe friction material must be uncontaminated and have plenty of life left. Brake fluid must be less than two years old, and kept that way by being regularly changed. It is vital that there are no leaks, cracks or corrosion affecting the brake pipes and check, of course, that calliper and drum wheel cylinder pistons are not seized or corroded.

It is important to note that post-1977 Midgets have a dual line brake system, so you'll need to bear this, and other differences to early cars, in mind when modifying the system.

The accompanying chart shows the modification options available.

Braking system modification options						
Car use	Fluid spec- ification	Metal braided hoses	Friction material front/rear	Servo	Bias valve	Disc size
Standard/ Mild road	DOT 4	Yes	Standard	Optional	No	Standard
Fast road/ Mild competition	DOT 5.1	Yes	Ferodo 3466F Mintex M1144	Yes	Optional	Consider large
Serious competition	DOT 5.1 AP550/ 600	Yes	Ferodo 4003F Mintex M1155	Optional	Yes	Large

BRAKE FADE

Reduced braking efficiency, long brake pedal travel and a strong burning smell are all symptoms of brake fade, which can be caused by overheating of the friction material or fluid, or both. When the brake fluid boils, symptoms will be a soft brake pedal (which sinks to the floor in extreme

FRICTION Vs TEMPERATURE Materials M1144 and M1155

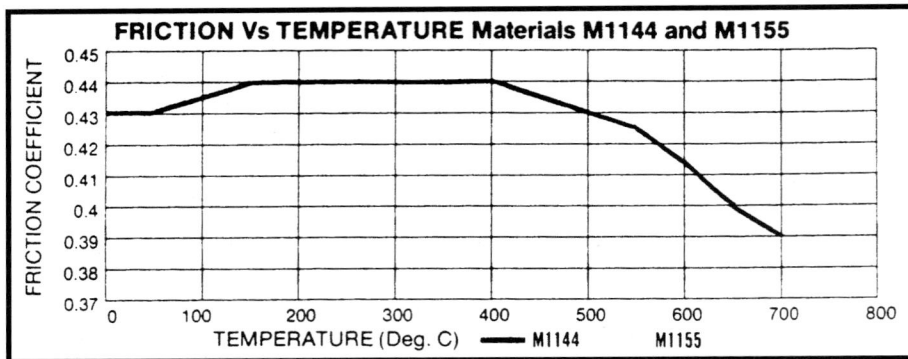

Friction *versus* temperature graph for Mintex brake friction material. *(Courtesy Mintex/Don).*

strips on the brake calliper and drums. Full instructions are included in the kits which, if used correctly, will provide you with vital information concerning the temperature range of your brakes.

A further point on brake temperatures is that it's necessary to get front and rear brakes operating within a fairly close temperature band, otherwise you may end up with brake balance problems.

Finally, when you have selected suitable friction material, be sure to follow the manufacturer's instructions on bedding it in before serious use!

cases), coupled with a loss in braking power. When the friction material overheats, the pedal will remain firm but there will be a loss in braking power. A strong burning smell does not mean that the friction material is fading, but it is close to overheating.

On a Midget/Sprite with uprated friction material, the most likely cause of brake fade is fluid overheating and a change to DOT 5.1 fluid or racing fluid, such as AP550, eliminates the problem.

FRICTION MATERIAL

Having decided that a change to a higher performance brake friction material is necessary for your tuned car, you need to consider your precise needs. Despite some advertising claims, brake lining material is split into two types: race and other. Although some braking material might just qualify as dual-purpose by being suitable for mild competition applications and road use. Competition braking material is not dual-purpose, but will give unrivalled performance when used as intended.

Mintex's newly-introduced asbestos-free brake pads have the following designations: M1144 replaces M171 and M1155 Supersedes M200. The optimum working temperature range of M1144 is 100 degrees C to 500 degrees C (212 to 932 degrees F). M1155 works as well as M1144 material from 100 degrees C (212 degrees F) and carries on working until it reaches just over 700 degrees C (1292

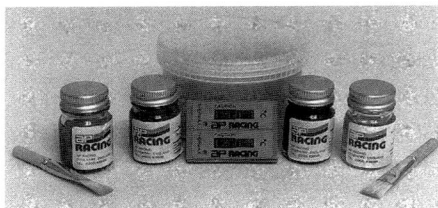

AP brake temperature recording kit.

degrees F). However, note that both materials work from cold. The M1144 material is better suited to road applications than the M1155 material since it has an edge in performance when cold and the M1155 material is slightly noisier. That said, if you really want to run M1155 on a road car there should be no problem in doing so.

Ferodo's asbestos-free range consists of a high performance road material and a race material. The road material is 3466F, described by Ferodo as having a maximum temperature of 500 degrees C (932 degrees F) with continuous and intermittent working temperatures of 225 and 300 degrees C (437 and 572 degrees F), respectively. The Ferodo race material carries the designation 4003F and has a maximum temperature of 750 degrees C (1382 degrees F) with continuous and intermittent temperatures of 225 and 400 degrees C (437 and 752 degrees F), respectively.

To discover the temperature range your brakes are working within, it's necessary to use thermal paint (available from suppliers like AP Racing) on the brake discs and temperature indicator

BRAKE FLUID

The brake fluid you use in your car's braking system is as important to braking performance as the brake components themselves. The best braking system money can buy will perform poorly if the brake fluid is not appropriate to the task in hand.

A modified road or racing car that makes more demands on its brakes than a standard car, will require a better brake fluid than that normally specified. This is because, as the brake components dissipate heat when in use, some of that heat is passed to the brake fluid. When the temperature of the fluid reaches boiling point, it vaporises. The vapour is compressible, resulting in the brake pedal compressing vapour instead of displacing fluid, the effect of which is a soft pedal and loss of braking power. The solution is to use a brake fluid with a higher boiling point.

Brake fluid types are mineral, silicone and polyglycol based. Mineral fluid can be disregarded since it is neither found in the Midget/Sprite or used in any competition application.

The principle difference between silicone and polyglycol fluid is that the former is not hygroscopic (it does not absorb moisture from the atmosphere). Silicone brake fluid will preserve the interior of hydraulic pipes and also has a long service life, but is expensive and unsuitable for most competition applica-

Brake fluid boiling (vaporisation) points				
	DOT 3	DOT 4	DOT 5.1	AP550
Dry	205 degrees C 401degrees F	230 degrees C 446 degrees F	268 degrees C 514 degrees F	290 degrees C 554 degrees F
Wet	140 degrees C 284 degrees F	155 degrees C 311 degrees F	191degrees C 375 degrees F	145 degrees C 293 degrees F

tions because it becomes compressible at high temperatures and does not have a 'dry' boiling point as high as racing fluids. Although silicone fluid is not hygroscopic, any moisture or water present or introduced remains 'free' water and can produce vapour lock at temperatures as low as 100 degrees C (212 F). Also, when changing from polyglycol to silicone fluid, or vice versa, you'll need to thoroughly clean the system and possibly renew the pipes and seals, though not everyone agrees on this point. The reason for this is that although silicone and polyglycol fluids are compatible, they will not mix. In the UK and USA there have been problems with brake seals when silicone fluid has been used in the Midget/Sprite.

All fluids are graded in relation to the way they are affected by temperature. There are four common standards: DOT 3 (J1703), DOT 4, DOT 5 (silicone) and DOT 5.1. The DOT value relates to the wet and dry boiling points of the fluid. The dry boiling point is measured using fresh fluid and the wet boiling point measured after the fluid has been exposed to a controlled humidity. The respective dry and wet boiling points for DOT 3, DOT 4, DOT 5.1 and some AP racing fluids are given in the accompanying chart.

It can be seen from the chart that a higher DOT number equates to higher dry and wet boiling points. The 'dry' standard is the important one because frequent changes mean the fluid never comes close to being classed as 'wet.' Fluid changes every 12 months, or less, and monthly for race cars using any fluid are recommended. Consult AP for fluid change intervals if you are using that company's racing fluid (because of its low 'wet'

boiling point it requires changing at more frequent intervals than conventional fluids).

Castrol Dot 4 and AP Dot 5.1 will give good service when used in the Midget/Sprite, whatever brake components are fitted.

Don't be tempted to use a race fluid in a road car because of its comparatively low wet boiling point. Put another way, a race fluid goes off (to the wet value) much more quickly than a normal fluid.

BRAKE HOSES

The fitting of metal braided brake hoses, such as the Goodridge stainless steel items, will give the brakes a good solid feel. Goodridge no longer supplies direct to the public but, instead, through a network of dealers.

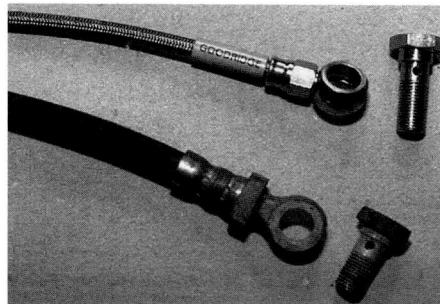

Standard flexible brake hose and Goodridge metal braided equivalent.

FRONT DRUM TO DISC CONVERSION

Various braking set-ups have been used on the Midget/Sprite and the earliest cars

had a non-servoed, single circuit drum/drum set-up. Later cars, from 1964, had a single circuit, non-servoed disc/drum arrangement. The very late cars had the same non-servoed disc/drum set-up but used a dual circuit brake line.

The disc/drum set-up is superior to the drum/drum set-up and it's possible to convert earlier cars to this specification by using the later car's components. The single circuit brake line layout is the easiest to use. Some drum/drum cars had a special factory conversion to disc brakes and wire spoked wheels for which, unfortunately, parts are no longer readily available.

Before you make the conversion from drum to disc, you'll need to decide whether you wish to use steel wheels or wire wheels since the hubs and discs are specific to each type of wheel. Remember, though, that if you convert the front wheels to wire wheels you'll also need to convert the back as well using special parts, which means changing the early back drums for later ones - more of which later.

The components required for the conversion are: stub axles and king pins, dust tubes and springs, hubs and studs, backplates, callipers, hose brackets, discs, pads and pad retaining plates with clips, new hoses with banjo bolts and copper washers, 3/4 bore master cylinder assembly (1098) or late type master cylinder assembly (1275). The conversion is made by removing all the drum parts and replacing them with the disc parts as per workshop manual instructions.

BIG BRAKE DISC KITS

There are several big disc conversions available for the Midget/Sprite although, if your car is wire-wheeled your choice is much more limited. A few kits use vented discs and some even use 4-pot brake callipers, but these conversions are very expensive. Bear in mind that improvement in front wheel braking efficiency should be balanced at the back, otherwise the result is a car that locks its front wheels.

Large front discs fitted to Simon Page's racing Frogeye.

Large disc conversion available from Mk Parts in Germany.

Large disc conversion on Ian Marr's car.

In Australia, Gillspeed retails a big brake disc conversion kit which uses Ford Cortina discs. At the time of writing, however, this kit is not suitable for cars with wire wheels.

The design of some of these kits is such that the front wheels are spaced outwards, which can cause clearance problems. There are two conversions recommended as being simpler and cheaper than the rest: in the UK the type 900 conversion from Peter May and in the USA the kit from Mini Mania.

Once a big disc kit has been fitted and the brakes settled in, it's recommended that, if your car has a brake bias valve, you experiment by increasing the rear brake loadings (having increased the braking force available at the front wheels, the ratio between front and rear will have been changed and this will, therefore, need to be restored). The ideal ratio will depend on the front to rear weight bias of your car and the specification of the braking system but can be established successfully by testing on private roads or the race track.

Peter May type 900 kit

The author was one of the first to fit a Peter May type 900 kit to a wire wheel Sprite and was very impressed with the kit and the results obtained. The kit is the same for bolt-on and wire wheel cars and uses a 9 inch disc as opposed to the standard 8.25 inch disc. The new disc is also quite a bit thicker than the standard item. Together, these two factors mean that the new disc has a greater capacity for absorbing and dispersing heat generated during braking. The calipers and disc pads

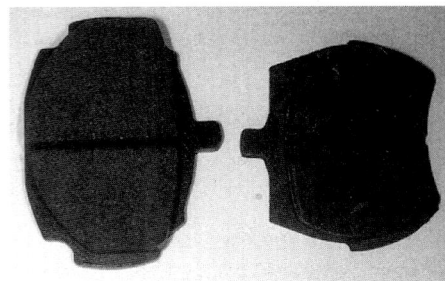

Peter May Engineering big disc conversion brake pad alongside standard size pad.

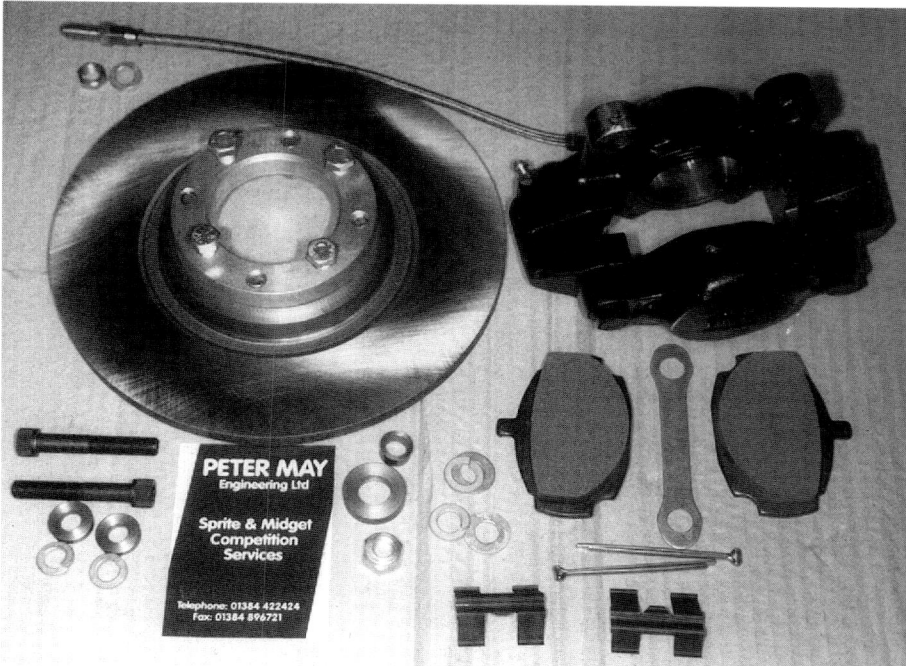

Everything in one kit for a large disc conversion from Peter May Engineering; this is just for one side. The kits are suitable for wire-wheeled cars, too.

Peter May 9 inch disc kit fitted to a Sprite.

around 25 per cent more than standard brakes. Peter May Engineering can supply the kit with the friction material of your choice. Note also that the kit comes with new braided steel brake hoses.

The kit contains an impressive set of fitting instructions that should be read thoroughly. A point worth noting is that, after installation of the kit, the steering will need to be re-aligned (tracked).

BEDDING-IN NEW DISCS

Although it might seem a good idea to use new pads with new discs this is not the case. The best thing to do is to run a set of old pads on new discs to condition the discs. Once everything has bedded-in, fit the new pads.

EARLY REAR DRUMS TO LATE REAR DRUMS

Early cars had single-acting piston rear brakes which can be converted to the double-acting piston rear brakes of later cars. If you're converting to later drums and use wire spoked wheels, you'll need halfshafts and hubs and hub extensions in addition to the parts required for the standard conversion. The parts required for the conversion are: backplates, drums, shoes and springs, cylinders and clips, handbrake link rod (specific to steel or wire wheels).

LARGE REAR DRUM CONVERSION

The standard rear drum brakes are more than adequate for most uses, and a brake bias switch will unload braking power to the rear brakes to prevent premature wheel lock-up. That said, if you are using the Peter May Engineering Ltd. type 900 conversion to larger front discs, you might want to fit bigger rear brakes to increase overall braking efficiency. Alternatively, if you are auto-testing or classic rallying, you might well find that you need a good

are larger than standard, again, allowing for better heat absorption and dispersion. The combined result is a front braking system that is considerably more resistant to fade than the standard braking system.

The benefits don't stop there, though, as the larger pads, calliper and disc give an increase in braking area of

A fully assembled 8 inch drum assembly. Note the cam adjuster which can be reached through a hole in the drum for easy adjustment.

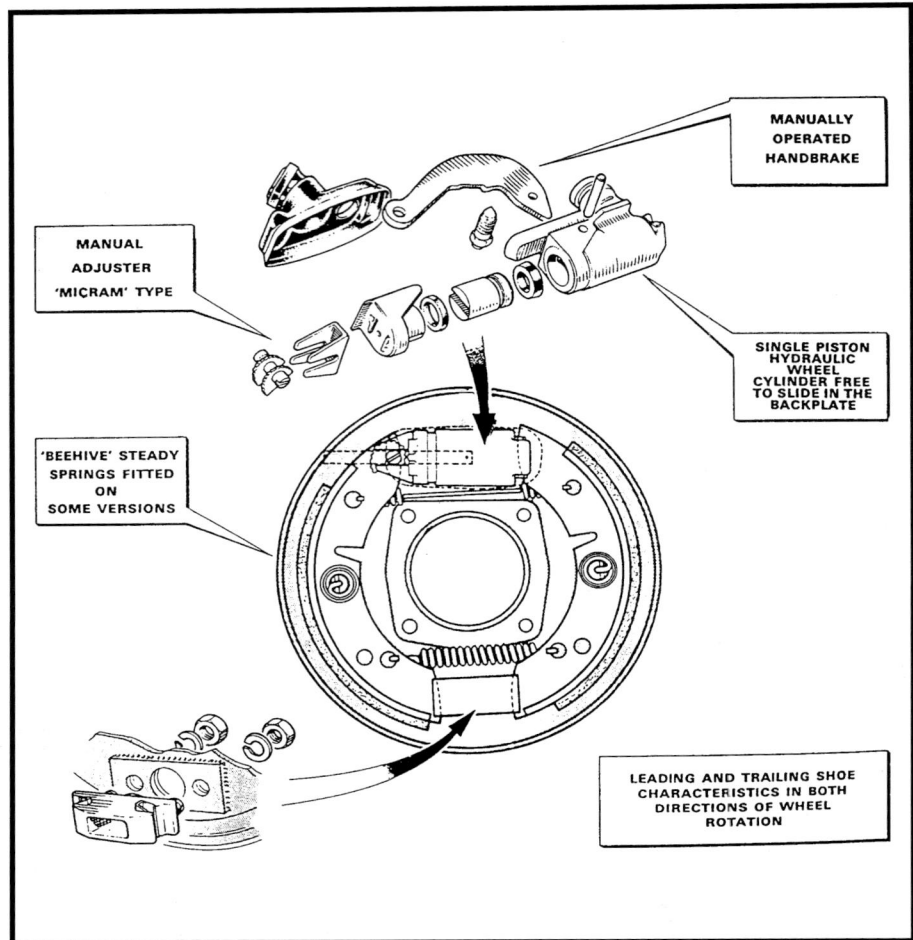

MANUALLY OPERATED HANDBRAKE

MANUAL ADJUSTER 'MICRAM' TYPE

SINGLE PISTON HYDRAULIC WHEEL CYLINDER FREE TO SLIDE IN THE BACKPLATE

'BEEHIVE' STEADY SPRINGS FITTED ON SOME VERSIONS

LEADING AND TRAILING SHOE CHARACTERISTICS IN BOTH DIRECTIONS OF WHEEL ROTATION

8 inch Lockheed drum brake assembly from Wolseley 1500.

handbrake action in order to handbrake-turn the car, in which case you'll find the standard rear brakes pretty inadequate unless the car is very light.

You can use Wolseley 1500 rear drums or the Riley 1.5 rear drums, both of which are of 8 inch diameter. Of these two, the Wolseley conversion is recommended since the Lockheed slave cylinders the Wolseley used are likely to be widely available, whilst the Girling items the Riley used are very scarce and doubtless will eventually become unobtainable (it is possible to use the smaller Frogeye slave cylinder with the Riley brakes which maintains brake balance front to rear and still provides an excellent handbrake). Replacement drums and shoes are not a problem since the late Morris Minor front drum brake shoes will fit. The company Wolseley 1500 Spares is a good source of parts for both brake sets and is listed in the suppliers section.

This is a very straightforward conversion, mostly comprising a swop from standard to 8 inch drums. The workshop manual instructions can be followed for removal of the standard backplate and brake assembly: note that a hub puller is required for this job. It's advisable to rebuild the slave cylinder on the new brakes before fitting and, possibly, fit a new cylinder as an extra precaution. The brake assembly will fit on the axle in any

one of four ways, but only one way will allow the handbrake linkage to be connected. The backplates are also handed and, again, if they are fitted to the wrong sides it will not be possible to connect the handbrake linkage.

The handbrake linkage should fit after a slight extra twist has been added (to ensure smooth operation and to ensure full release: heat the rod to make the twist permanent). You may find that, even with the handbrake fully adjusted, lever travel is excessive which is most likely to be a result of the drums having been skimmed once if not twice, and thereby creating extra travel. Rather than purchase new drums, you can shorten the handbrake rods by cutting each rod in two, losing about

0.25in (6.35mm) in length in the process. Next, have the two halves welded together, sleeving the rods with steel tubes to make each rod strong at the joint (it's useful to note the correct orientation of the ends and to allow for the extra twist required: however, if you get it wrong, the rod can be heated and twisted again using a welding torch). Alternatively, if you have a steel wheel axle, you could use wire wheel axle handbrake rods which are shorter by just the right amount.

Despite the Wolseley slave cylinder being of the single piston sliding action type, it's not necessary for it to have a flexible hose connection, a point confirmed by AP Lockheed.

The author had his car's rear brakes

and handbrake roller brake tested before carrying out this conversion. The brake readings for the rear wheels and standard brakes were 50 each side and 50 when used together: a very low figure due to the brake bias valve setting. The handbrake reading was 150 each side and 150 when used together. The brake readings for the 8 inch drums, once the linings had bedded in, was significantly higher than the standard readings both for the rear brake action and the handbrake. As a measure of efficiency against the car's weight there was a gain of 12 per cent, sufficient to allow a handbrake turn to be made on dry tarmac or concrete.

After these tests, the bias valve was progressively adjusted to increase the rear brake loading and to achieve higher overall braking efficiency. Even as little as half a turn was excessive, causing the rear brakes to lock before the front. This all changed, though, when a Peter May type 900 conversion was fitted to the front brakes, increasing front braking force by an average of 25 per cent with similar percentage gains in overall braking efficiency once the bias valve was appropriately adjusted.

ALLOY REAR BRAKE DRUMS

Available from K.A.D in the UK and Mini Mania and others in the USA, are aluminium alloy rear brake drums. Originally designed for the Mini - hence the common name "Minifins" - they are replacement brake drums with fins to aid cooling. The advantage in using them is thermal efficiency and overall (and unsprung) weight saving.

BRAKE SERVO

An improvement in braking is possible by fitting a brake servo. A servo kit for the Midget/Sprite is fairly expensive, but is money well spent. A servo kit should include the following parts: servo, mounting brackets, vacuum pipe and additional brake piping The AP kit (part

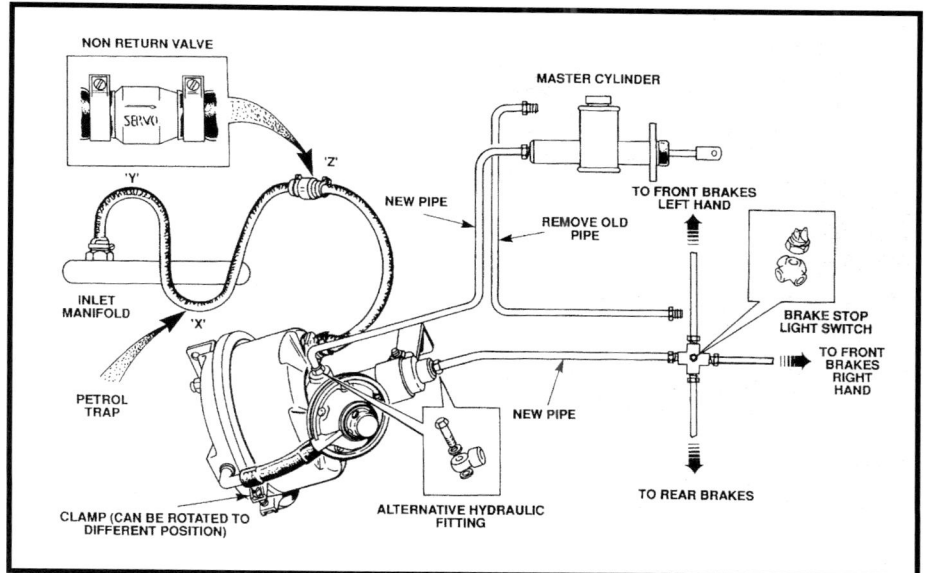

The AP brake servo kit is easy to fit and this drawing illustrates how the hydraulic piping is modified to put the servo in series between master cylinder and union. *(Courtesy AP & redrawn by Dave Robinson/Sharon Monroe).*

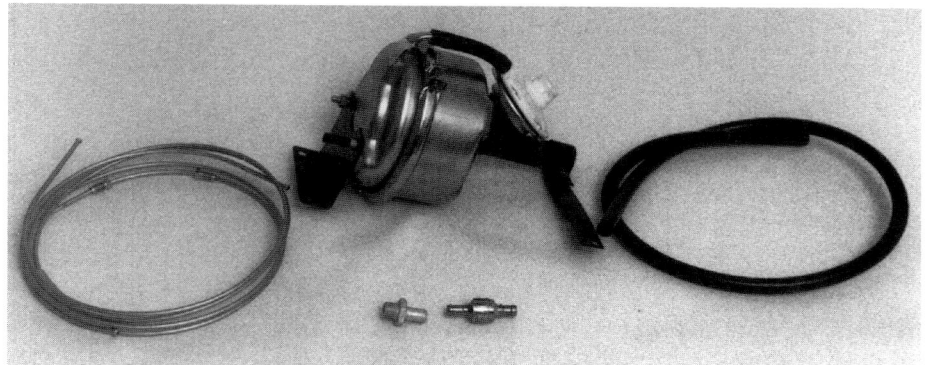

AP brake servo kit complete ready to fit to a Midget/Sprite.

no. LE 72696 pre-1977 cars) contains all these and a selection of fittings. The servo for the Midget/Sprite is different to most car brake servos in that it is separate from the master cylinder; for this reason it is known as a 'remote' type. The old Mini Cooper S had a remote servo and the servo in most kits is the same type and size as the Cooper's. Anything bigger than this is too large. Mini Spares of London can supply the servo and most of the parts and so can AP, the manufacturer of the kit.

A servo works by using vacuum from the inlet manifold, affecting one side of a

large diaphragm to boost the hydraulic pressure. This in itself does not improve the car's braking capability but it does reduce the amount of pedal pressure needed. This extra pressure from the servo allows full utilization of the car's braking potential as maximum braking is achieved at the point just before wheel lock-up. Of course, if you're a budding Mr/Ms Universe don't bother with this mod; however, since most of us don't fit into that category then, as tyre width or gripping properties increase, it will become even more difficult to achieve a near lock-up

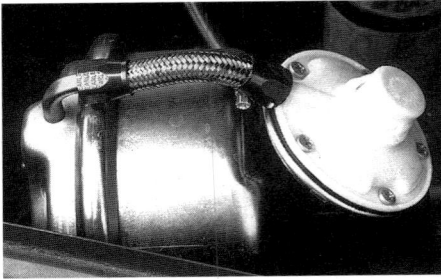

The servo unit should be positioned under the wiper motor. Bonnet clearance is tight.

Braided metal hose for servo vacuum take-off.

condition without a servo. An interesting point is that extremely few race car drivers use servo-assisted brakes as they prefer a solid pedal and, inevitably, the servo adds a little sponginess to the feel of the brakes.

A servo fitting kit should contain full fitting instructions but, in case it doesn't, and for the benefit of readers who buy the components separately, a basic fitting procedure follows.

Fit the servo unit first (you can purchase brackets or fabricate your own from mild steel or aluminium sheet metal or strip). On a right-hand drive, A-series-engined car, the best place to fit the servo is just below the wiper motor, with the narrow end pointing away from the front of the car. If you have a 1500 Midget, just

below the wiper motor is the best site for the servo (this will require fitting a smaller washer fluid reservoir and moving the ignition coil). A different location may be necessary on left-hand cars; the main objective is to try and keep all the brake and vacuum lines as short as possible. Wherever you site the servo, be sure to allow enough room for the bonnet to close.

With the servo unit secured in place, fit the vacuum pipe from the unit to the inlet manifold. Hobbsport can supply metal braided hose and associated fittings suitable for servo/manifold vacuum pipe, as well as cheaper rubber hose. The braided hose is desirable because it is much less likely to chafe and leak. If you are using a second-hand servo system, it's worth ensuring that it has the non-return valve-type manifold union which allows a

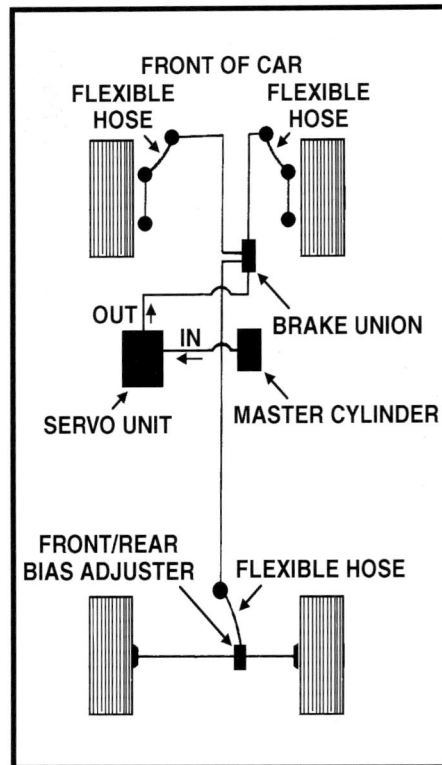

This simple drawing illustrates the brake pipe routes when both a servo and a brake bias adjuster are fitted: both are marked. This drawing represents single brake line cars. (Courtesy Dave Robinson/Sharon Monroe).

vacuum to be stored for periods of non-operation - new AP units have this valve built in.

Most SU manifolds, standard or competition, will have provision for a servo vacuum pipe take-off and will only require the blanking screw's removal and substitution by a hose union. Weber manifolds may not have this provision so the manifold will need to be removed - otherwise swarf will be sucked into the engine - for drilling and tapping.

The last job will be to fit the hydraulic piping. The existing rigid piping from the brake master cylinder to the four-way brake union is removed. A new rigid pipe is run from the master cylinder to the servo hydraulic inlet. Note that all servos have an inlet and outlet; be sure you fit them the correct way round. Another new rigid pipe is used to run from the servo outlet to the four-way union. When you purchase your pipe ask for the copper type: it's a minimal extra expense, but much greater corrosion resistance makes copper a preferable choice. If you wish, you can buy a brake pipe flaring kit and make all your own pipes.

The braking system will require bleeding before use. Keep a close eye on the fluid level when bleeding as you have a lot of new pipe to fill with fluid. The quickest way to get air out of the system is to bleed the servo outlet pipe just before it's screwed into the four-way union; then connect it and bleed the rest of the system in the normal manner.

BRAKE BIAS VALVE

Standard brakes which are properly adjusted and in good order should be able to lock the wheels of the car. However, because of the Midget's/Sprite's light back end, and the weight transfer from the rear of the car to the front during braking, the rear wheels may lock first, starting a tail slide which can result in a spin (depending on the speed at which the slide occurs). To correct this problem a bias valve needs to be added to the brake hydraulic system.

If you've fitted a big disc conversion to the front brakes, they may now lock

Mini Mania's twin master cylinder set-up with remote fluid reservoirs. *(Courtesy Mini Mania).*

Two close-ups of the twin master cylinder set-up on Dave Grove's Frogeye.

Twin master cylinder brake balance on Simon Page's Frogeye.

before those at the rear. The way to achieve optimum braking (front wheels lock a fraction of a second before the rear) is, again, to fit a bias valve in the hydraulic system.

Once fitted, a bias valve can be adjusted to split hydraulic pressure between front and rear brakes until an ideal balance is achieved. A suitable Lockheed bias valve is available from Ripspeed Ltd. and there are other, more expensive, types available which have pre-set selector positions (available from Goodridge hose dealers).

To fit a bias valve, disconnect the three brake pipes from the three-way union which is mounted on the differential (axle) housing. Remove the union and replace with the bias valve. The brake line from the front of the car can be fitted to the valve. The pipes which connect the new bias valve to the rear drums have to be completely removed so that they can be re-flared to fit the bias valve unions. The type of flair that is required for the Lockheed valve is concave.

Once the valve is in place and the

pipes have been refitted, the system is bled in the usual manner. The brake balance from front to rear is controlled by turning a key in the valve.

An alternative, more expensive, solution to achieving brake balance is to fit dual master cylinders with an adjustable brake balance bar. The difficulty, however, with completing this job on the Midget/Sprite is that, unlike most master cylinders, the Midget/Sprite master is set at an angle. This necessitates the use of racing master cylinders and the extra expense and complexity means it's a job for the experienced mechanic only.

Chapter 12
Instruments & Electrical Components

OIL TEMPERATURE GAUGE

Even if you've fitted an oil cooler and, perhaps, a thermostat too, your engine can still overheat. For instance, the oil thermostat could stick in the closed position. The fitting of an oil temperature gauge is a wise precaution for cars with high performance engines.

Gauges can be operated by two types of sensor. One type fits into an oil line, but the oil is always likely to be cooler in the line than in the sump. The second - and much more common - type fits in the sump.

Normally, sump fitting sensors can only be fitted with the oil drained and the sump removed (though you might find one that fits into the drain plug). Therefore, the most economic way to do the job is to wait until the next oil and filter change is due (especially if you use synthetic oil which is too expensive to waste). The parts required to plumb in the gauge should all be contained in one kit; the only exceptions might be a sump gasket and a supplementary instrument pod.

Fit the gauge and its mounting pod in

On this car, oil gauges have prominence and petrol and water gauges sit in a small pod under the dash panel.

the desired location: if you're short of ideas, how about below the existing dash? (Check pod-to-gearlever clearance in first and top gear positions). If the unit you are fitting is a capillary-type gauge, it's necessary to pass the sensor and capillary tubing through the engine bulkhead before fitting to the sump (check to see if an existing grommeted hole is usable).

When the sensor is dangling in the engine bay check clearance to the chassis leg. Be sure to mark a sensor location that allows adequate room for the sensor (in other words, check how far it protrudes).

This (arrowed) is where the oil temperature sensor fits in the sump in the author's car (it would have been better placed nearer the drain plug for ease of removal).

Sensor for oil temperature gauge.

Drain the oil, remove the sump and drill the hole in the required position. The sensor mount must now be brazed into the hole. After studiously cleaning all metal swarf from the sump, fill with paraffin and check there are no leaks. If everything is okay, drain and clean the sump. Install the sensor and, using a new gasket, refit the sump and drain plug and refill the engine with oil. Check for leaks before and after the engine has warmed up.

Now that you have an oil temperature gauge, you'll notice that the engine can take quite a time to reach operating temperatures. If no temperature indication is seen on the gauge, re-check the installation. You'll quickly gain experience of the temperature your engine is likely to run at, be it around town, on the motorway or on the racetrack. Use this experience to determine if your engine requires a larger, or maybe even smaller, oil cooler.

OIL PRESSURE WARNING LIGHT/ BUZZER

The standard oil warning light on your Midget/Sprite (where fitted) is operated by an oil pressure switch in the cylinder block, which may also operate the oil pressure gauge if it is the electrical type (1500 Midget). Typically, the switch operates at between 3.5psi and 7psi (24.13 and 48.26Kpa) depending on the switch used, although it may be fractionally more or less than this. As a means of indicating that oil pressure has been achieved this device is fine, but it's next to useless if a sudden loss of engine oil pressure occurs at high engine speed as, with only 7psi or less of oil pressure, you may have the beginnings of terminal engine failure.

To provide earlier warning of low oil pressure replace the standard switch (where fitted) with one that trips at the higher pressure of 20, 22 or 25psi (138,

Adapter to take oil warning light (25psi switching) sensor and oil pressure gauge take-off.

Simon Page's car's dash is neat and functional and includes turbo boost gauges. Note large low oil pressure warning light in centre of panel and Pool ball gearlever knob.

151.6 or 174.4kpa). For 1500 Midgets this switch is part number TT998, available from any Midget/Sprite performance parts supplier. For A-series-engined cars that don't have an oil warning light and a non-electric oil pressure gauge, you'll need a union fitting with an additional outlet for the pressure switch to replace the standard gauge fitting. Speedograph Richfield instrument accessory code TP is the part you require.

A 22psi (151.7Kpa) oil warning light switch is available from Mini Spares, part number HPS1. For cars with the non-electrical oil pressure gauge, it's possible to replace the standard fit copper pipe and hose connection from union to gauge with braided steel hose measured to fit your precise requirement. Any accessory shop can sell you an amber pilot light which is connected in a simple circuit to the pressure switch. For a racing car it is a good idea to replace the standard low oil pressure warning pilot light with a much bigger, brighter unit. It's also possible to wire in a warning buzzer that will activate simultaneously with the light.

WATER TEMPERATURE WARNING LIGHT/BUZZER

The car's standard water temperature gauge will give a reasonable indication of water temperature.

However, if you don't watch the gauge like a hawk, the first indication that a problem exists may be an embarrassing cloud of steam from a burst hose or radiator overflow which you may not be aware of until the resultant excessive engine heat has already wreaked havoc.

The solution is to fit a simple temperature sensing switch (available from Kenlowe) which will activate a dashboard-mounted idiot light when the coolant temperature reaches an adjustable, pre-determined level. As an alternative, or in addition, a warning buzzer can also be wired to activate simultaneously with the light.

SHIFT LIGHT

If you have a Stack tachometer you have a built-in shift (gearchange) light function. However, if you're using another make of tachometer, you can purchase a separate shift light.

Cockpit-mounted speed shift unit from Armteca. (Courtesy Armteca).

The Armteca shift light is called "Speed Shift" and incorporates a rev limiter and downshift light function, as well as enabling full throttle gearchanges to be made without lifting off the throttle or blowing up the engine. The speed shift is only recommended for use on racing cars, but the law does not prevent full throttle upshifts on the public highway.

The unit is easy to fit and comes with full instructions.

REPLACEMENT TACHOMETER

On high performance cars one of the most important instruments is the tachometer (rev-counter), so you might wish to replace the standard item with a more accurate instrument. The standard tachometer can often be inaccurate before 6500rpm let alone beyond.

If you are after the ultimate in aftermarket tachometers, take a look at the range from Stack. Not only are these highly accurate instruments, but there is a wide range of dials to suit any application. In an independent evaluation of tachometers by a British car magazine, the Stack rev counter was the only unit which was 100% accurate.

The Stack unit comes complete with extremely comprehensive fitting and usage instructions. Two switches are required to go with the unit (normally supplied with the kit). These switches control the telltale maximum display and reset the telltale

Stack tachos are available in a range of faces.

Stack tacho mounted in place of original instrument.

facility. In addition, not only can you have a tachometer telltale, but also an action replay facility of up to five minutes. Better still is a longer performance analysis with up to 25 minutes' recall on a printout.

Stack tachometers come in two sizes 3.15in (80mm) or 4.57in (116mm). However, the 3.15in (80mm) hole required for the small unit, is smaller than that of the standard tachometer. Therefore, a blanking plate needs to be fabricated to take up the difference between the old and new sizes. Cut a hole in a simple flat sheet of aluminium (try a local car body repair shop for this if you don't have the necessary tools) and then rivet the plate in, after which it can be painted to match the dash panel. The hole in the blanking plate can be absolutely central, but it might be better to have the new hole eccentric to the old hole in order to maximise the driver's view of the instrument.

The Stack tachometer has a block connector that fits into the unit at one end and to respective wires at the other. It will be necessary to discard the old tachometer wiring and start afresh for the Stack unit. Note that the Stack unit is self-illuminating, so does not require the instrument light fittings that were used on the old tachometer.

All aftermarket tachometers come with comprehensive fitting instructions and are simple to fit. However, you may need to fit dash-mounted warning lights to replace those previously situated in the face of the original tachometer.

Mechanical tachometer kit

An alternative to the conventional electronic tacho is to use a mechanical item. Mini Spares Ltd. retail an 3.15ins (80mm) kit which is driven from the end of the camshaft, via the timing gear case. The kit can be used with the standard timing gear/case or with the Mini Spares belt drive timing gear/case. Turbocharged Frogeye racer Simon Page uses such a tacho on his race car and has found the unit to provide excellent service.

SPEEDOMETER MODIFICATIONS

If your car has a highly tuned engine, a 5-speed gearbox conversion, an overdrive conversion, or has had a change in rear axle ratio, you may have problems with speedometer accuracy.

The standard Midget/Sprite speedo is fine for normal use but, with a tuned roadgoing car it's preferable to know exactly what speed you're travelling.

A speedometer can be checked for accuracy on the rolling road and the results noted. However, it's not practical to look at a set of corrected figures whenever you approach the legal speed limit. A change in differential ratio from standard will render the speedometer wildly inaccurate, reading on the slow side for higher gearing and the fast side for lower gearing - neither of much use.

One solution would be to purchase an aftermarket speedometer, but they are likely to lack a main beam warning light and not have useful calibration - even the above-average Midget/Sprite is unlikely to be capable of 160mph! I suggest the simplest way forward is to purchase a second-hand 120mph-face MGB speedo and send this, and your Midget/Sprite unit, to Speedograph Richfield, who will build you a hybrid of the two. Actually, it's only necessary to send the old unit if the cases are different, which can sometimes

Hybrid speedo from Speedograph Richfield - calibrated to suit a particular application.

Mark wheel and bodywork to measure wheel revolutions in relation to speedometer cable revolutions.

happen. The 120mph-face MGB units come in two main types: those from overdrive cars and those from non-overdrive cars. Either type will suffice for conversion but it is helpful for Speedograph to know which type it is where possible. Late 1500 Midget speedometers also read to 120mph, but will not match the face of the standard car's rev counter and will need recalibrating since the 1500 uses a different rear axle ratio and a different gearbox.

In order to rebuild the MGB unit to produce a Midget/Sprite speedo with a 120mph face that is absolutely accurate, Speedograph requires that you undertake the following -

1. Disconnect speedometer flexible drive at the instrument end.
2. Jack-up one driving wheel and support the car with an axle stand. (Let the company know if your car is fitted with a limited slip differential).
3. Mark suspended driving wheel with a chalk line or masking tape.
4. Make a corresponding mark on the rear wing.
5. Make a small arrow from light cardboard and press it onto the end of the speedometer drive inner cable.
6. Revolve by hand the driving wheel exactly twenty times whilst an assistant

counts the number of revolutions the inner cable makes (to the nearest one eighth of a revolution).
7. Note the make and size of the tyre and wheel revolution per mile figure, if you have it.
8. Send the information from points 6 and 7 to Speedograph, along with the name of the speedometer make (usually Smiths) and part number, if known. Also give a brief explanation of what your requirement is.

Alternatively, Speedograph can manufacture and supply a speedometer to suit any particular style or requirement. Another service on offer is that of rebuilding a speedometer from kph to mph, or vice versa. The Toyota or Ford 5-speed gearbox conversion presents no problems and, as a cable manufacturer, Speedograph can reproduce at any length a speedo cable for your gearbox/speedo combination.

INSTRUMENT ILLUMINATION

When wiring any supplementary or non-standard instrument light, you'll generally find that the bulb live wire is red in colour. The standard wiring colour for instrument lights on the Midget/Sprite is red and white. In order to have matching wiring it is necessary to remove the bulb holder and the bulb from the instrument and heat the connection with a soldering iron to remove the red (or other colour) wire. The correctly colour-coded red and white wire can be bought from companies like Merv Plastics Ltd., and soldered into the connection as a direct replacement.

The easiest way to connect the illumination wiring of supplementary instruments is to fit the bare end with a bullet connector. Then cut an existing instrument light wire in two and solder, or crimp, a bullet connector on each end. The feed and two instrument wires can then all be inserted into a two-bullet or three-bullet connector block to complete the circuit.

HEADLIGHT UPGRADE

If you are doing any amount of night driving, you might wish to upgrade the lights of your Midget/Sprite. The most obvious and easiest way to do this is by fitting spotlights. However, since any spotlights should only operate in conjunction with headlight main beams, you'll be no better off whatsoever when driving on dipped lights. The solution is to upgrade the existing headlights by fitting quartz halogen units.

On the face of it, replacement of standard 50/40 (Mk1 Sprite) or 60/45 headlamp bulbs with similar wattage quartz bulbs seems to represent no increase in power and hence light. However, a modern halogen bulb produces about 70 per cent more luminous flux (light) than a conventional bulb.

All Midgets/Sprites use normal 7 inch (178mm) sealed beam units which are easily converted to halogen. Halogen bulb units are a direct replacement for existing bulbs, except in the case of Frogeye Sprites which use the old style pre-focus bulbs. For Frogeyes, the bulb harness connector block will need to be changed and Hella Ltd. can supply the units for right-hand-drive cars under the part number 1L6 002 395-261 (less bulbs).

If you are looking for even brighter lighting, it's possible to find halogen bulbs of 100/80, 130/90 or 160/100 ratings, but check local legislation regarding use of such bulbs before fitting them. Two problems need tackling before using such bulbs. The first is that high wattage bulbs can double the loading on switches and, to overcome this, a relay must be fitted. The second is that the standard wiring is not designed for use with high wattage bulbs, and must therefore be replaced with wiring rated for a higher current.

Depending on the wattage of your chosen bulbs, you may need a relay for both the main beam and dip beam circuits (the safe limit for switching without a relay is 65 watts). Each relay must be wired through a fuse to the battery main feed or through the fuse box. To avoid a spaghetti-like wiring maze from the live

INSTRUMENTS & ELECTRICAL COMPONENTS

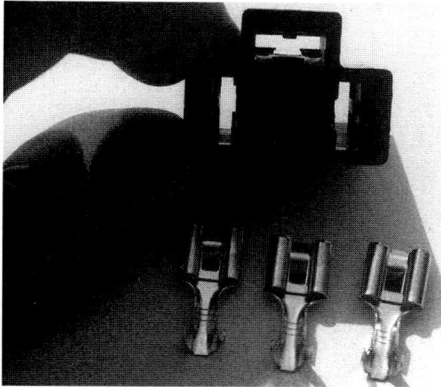

Connector block and spade connectors.

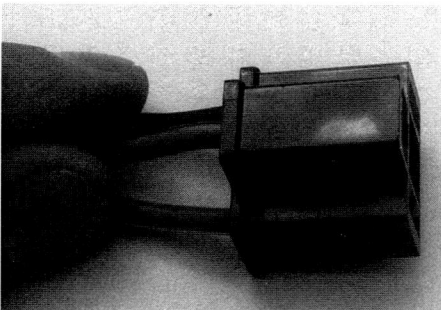

Headlight main/dip beam connector block wired-up with heavy duty wiring to cope with 160W main beam/100W dipped beam.

terminal of the battery, it's recommended the relay(s) is wired through the fuse box. However, as the Midget/Sprite fuse box has only two outlets (four on USA models), it's preferable to fit a supplementary fuse box with a larger number of outlets rather than use piggy-back-type connectors. Fuse boxes are available with four, six and eight outlets and you may wish to use the eight outlet box to allow for fitting of accessories in the future. The later type Mini fuse box can also be used.

If you stick with the standard fuse box, make sure that the main feed, via the relay for the headlamp, is not across the fuse in the fuse box but on the 'feed' side. Then ensure that there is a fuse link in that headlamp feed with an appropriate size fuse, or else the lights go out at night.

To determine the correct rating for headlamp wires, divide the bulb wattage by the voltage to get an amps rating. For instance, using 100 watt 12 volt bulbs as

an example, 100 is divided by 12 to get 8.34amps, the nearest cable rating (erring on oversize for safety) is rated at 8.75amps. Standard cabling on the Midget/Sprite is 8amps from which it can be seen that standard wiring is marginally undersized for 100 watt bulbs. 160 watt bulbs require 13.34amps and the nearest cable size is rated at 17.5amps. To determine the load rating for the relay - which will be required for bulbs over 100 watts - add the amps figure for each light (eg: for two 100 watt bulbs this will be 8.34 plus 8.34 equals 16.68amps.

Where a bullet connector is of insufficient size for the cable (only a problem on the thickest cables), the bullet can be drilled out to a larger size. It is strongly recommended that all bullet connectors are both soldered and crimped.

All the wiring you are likely to require (in the correct colour codes), as well as relays and bullet connectors, etc., can be obtained from specialists like Merv Plastics Ltd.

BATTERY

If you have a modified car you might want to fit an uprated battery to ensure that, even in the coldest weather, there's sufficient cranking power to start the engine. Alternatively, you might want to fit a small, lightweight sealed battery for racing only. This section will lead you through what is available for various applications, considering both power and physical dimensions.

Although there are numerous battery manufacturers to keep things simple, only Lucas products will be mentioned here (any of the major battery manufacturers will be able to supply equivalent batteries under their own brand names). The modern equivalent of the original equipment battery is the Lucas "038" or, for a higher specification, "015," both of which are detailed in the accompanying chart along with batteries of different dimensions which will fit without any modifications to the battery tray. An important point is that the physical size of any battery is not

Lucas positive (+) earth battery choices*			
	Standard	Heavy duty	Very heavy duty
Lucas part number	015	004	074
Dimensions			
Length	9.4ins (338mm)	7.9ins (200mm)	10.1ins (255mm)
Width	5.3ins (133mm)	6.7ins (170mm)	6.9ins (175mm)
Height	8.1ins (205mm)	8.9ins (225mm)	8.1ins (205mm)
Cell type & orientation	round post	round post	round post
Output Cold cranking (amps) SAE	335	380	470
Ampere capacity	40	50	66
Weight (wet)	11.5kg	14.0kg	17.6kg

For negative earth cars (which have the posts in the opposite positions) and 1500 Midgets (which have the posts furthest from, rather than closest to, the bulkhead) the equivalent batteries are: 016, 077 and 092.

necessarily a guide to how powerful it is.

For racing cars, where a sealed battery is required, a battery from the DMS Technologies Varley red top range is likely to be a good choice. Note that, if your car is going to be used in motorsport, it's usually a requirement that the earth lead is yellow. The easiest way to achieve this is by wrapping yellow insulating tape around it.

INTERIOR DIPPING MIRROR

A useful modification to any early car is to exchange the standard interior mirror for a dipping mirror which will make night driving easier by cutting out the dazzle from the headlights of the car behind. For the very early cars, the standard dash-mounted mirror can be replaced with a stick-on dipping mirror available from accessory shops. For later cars, the neatest solution is to fit the dipping mirror from a 1500 Midget.

To fit the mirror you'll need the whole assembly that attaches to the screen frame, since the new mirror will not fit onto the old mount. Fitting will require removal - by drilling out the rivets - of the old top plate, after which the new top plate can be riveted on. The rest of the dipping mirror assembly is a direct replacement of the old type.

TWO-SPEED WINDSCREEN WIPER CONVERSION

For anyone wishing to convert their Midget/Sprite wipers from single speed to two speed there is a simple and cost-effective means of doing so. The conversion can be undertaken using either a new or second-hand wiper motor.

The conversion requires a Lucas round-type two-speed motor. Cars which used this type of motor include Morris Marina, Austin Maxi and Mini. The Lucas part number for the Austin Maxi two-speed wiper motor is new 75664, exchange LRW110. Of the three , as will be clearer later, the Marina or Mini motors are

preferable. If your car uses the earlier three-stud fitting motor, you cannot swop motors without using the round-type motor complete with the later drive cable and wheelboxes.

As the wiper motor has a running current of only 3.1amps, the new wire required for the second speed need only be of 0.01in (0.33mm) thickness or more; the new wire can be taped to the existing wiring loom. However, it's preferable to purchase new wires of the correct colours along with bayonet fitting connectors and plug these into the old connector block. A three-way switch will be needed to complete the conversion. More of this later.

To match the existing dash layout, you can use a Midget/Sprite three-way headlight switch to replace the old two-way wiper switch. Depending on the year/mark of your car, you may adopt a different switch to match the existing switch gear.

Before the replacement motor can be fitted the old one must be removed. First, disconnect the wiring block connector from the motor and unfasten the motor brackets. Remove the motor drive cover to expose the drive pinion and cable. The drive cable can now be pulled off the pinion crank and the motor removed from the car.

At this point it may be worth checking to see how freely the wiper wheelboxes turn independently of the motor by pulling the wiper cable to and fro. If there is a great deal of resistance it's most likely due to excessive friction in the wheelboxes, or a kinked drive cable or cable tube. Removal of the cable will reveal if the problem is with the cable and tube or the wheelboxes. If it's the wheelboxes, remove them and drill a small hole in the body of each. Through this hole, squirt a silicone-based lubricant such as WD40. Turn the wiper blade shaft in the wheelbox and repeatedly squirt the lubricant into the unit until it frees.

The other cause of stiff wiper operation - excessive friction between the control cable and the control tube - can also be easily solved. If the control cable is

Right: Wiper motor with cover removed - large pinion has degree of sweep stamped on it: 110 degrees in this case.

kinked it will need to be replaced and if the control tube is damaged it, too, will need to be replaced. However, the control tube is not available ready to fit in two parts as a direct replacement but comes as a single tube which will require cutting and flaring. The bore of the tube is such that your local garage may not have a flare large enough to do the job, in which case, try your local commercial vehicle garage which uses larger bore piping on a regular basis. The flare shape is not critical and is only required to hold the pipe in position in the wheelbox.

A better alternative to ordering the pipe from an MG parts supplier is to use small bore copper central heating pipe, which is a tenth of the price. I recommend 10mm (0.39in) bore pipe which is larger than the quarter bore standard steel pipe. The advantage (aside from cheapness) is that the central heating pipe is much softer and therefore easier to bend. The increase in bore size also makes it much less sensitive to tight radius bends. Finally, if, once your wipers have been reassembled, you find that they sweep in opposite directions, it means you've fitted one of the wiper wheelboxes upside down!

The single-speed Midget/Sprite wiper motor uses a drive pinion that produces a sweep of 110 degrees; the Austin Maxi motor produces a sweep of 90 degrees and the Morris Marina a sweep of 120 degrees. Minis are either 110 or 120 so, in order that the sweep remains the same after conversion to a two-speed motor, it will be necessary to swop the drive pinion.

To do this, put the two-speed wiper motor alongside the single speed motor and remove the drive motor covers. Turn each motor drive pinion face down and release the circlip from the centre of the pinion shaft which protrudes from the back cover. This will release the pinion which can now be removed. By swopping over the two drive pinions the two-speed motor now has the correct sweep stroke for the Midget/Sprite. Reassembly is simply a reversal of strip-down.

If you use the Morris Marina motor, or the 120 degree Mini motor, the 120 degree sweep drive pinion can be retained and the new motor now not only produces a second speed, but a useful additional 10 degrees sweep as well. The drive pinion can be purchased from any Lucas parts supplier under the part number WGB 227. In the USA, Seven Enterprises stock the 110 and 120 gear and two speed motors.

Having sorted the replacement motor, fit it to the car and connect the electrical block connector. Run an extra wire to the spare spade terminal on the two-speed motor or, if you prefer, use the connector block which came from the same car as the motor along with the wires which should follow the same colour coding. If you have mixed up the wires, re-attach them according to the following guide (noting that each pin on the connector block that attaches to the wiring loom has a number): Terminal (T) 1 black, T2 brown/green, T3 blue/green, T4 green, T5 red/green. Run the black wire to a suitable earth. Run the green wire into an ignition live in the car's main loom. Taking a standard headlamp switch with numbered connections, connect as follows: T3/4 to red/green from the wiper connector block, T5 take a green from an ignition feed from the main loom of the car, T7 to blue/green, T8 to brown green. If you test the switch you'll find that one position is off, the middle position is slow wipe and other position is fast wipe. However, when you come to install the switch into the dash, you'll find that it operates in the opposite fashion to the other switches on the dash - in other words, up is on rather than down is on. The simple remedy is to file a flat on the switch and fit it in such a way that, with the toggle on, the switch is off. Note that the motor should still correctly park the wipers, though the wiper arms may need to be taken off and reset if you are using a motor with greater sweep than standard. If you're using a later type, or completely different switch, you'll find the use of a circuit tester helpful in establishing the correct contacts.

Chapter 13
Preparation For Motorsport

INTRODUCTION

It's not possible to give totally comprehensive instructions for preparing your Midget/Sprite as regulations vary not only for different types of motorsport, but also for different series. The regulations you will need to refer to will be available from the sport's or series governing body in your country.

What this chapter will do, however, is to point you in the right direction with regard to basic preparation and the modifications which are generally mandatory in all classes for competing with a Midget/Sprite. Even if you do not want to compete in motorsports, you may want to upgrade certain aspects of your car to make it safer in the event of an accident: for instance, fitting a roll bar, full harness or fire extinguisher.

COMPETITION SEATS

If your car has early type seats, the easiest and the cheapest way to upgrade the seats is to obtain seats with headrests from a

Corbeau Sprint seat fitted to the author's Sprite; note proximity to roll bar and door pillar.

later car. The headrest will help prevent a whiplash injury to the neck in the event of an accident. Swopping from early to late seats is very simple: a workshop manual will give you all the advice you need.

If you are fitting a sports or racing seat, the main thing to consider is whether or not you will be using it in conjunction with a full race harness. If you are, be sure to choose a seat with harness slots. There is a wide selection of sports seats and these can be purchased by mail order from the manufacturer or a local tuning parts shop. Be sure to check the seat measurements before purchase, though, because not all seats will fit the Midget/Sprite. The problem of limited space will be further exacerbated if a roll bar is fitted. The author has experience of three seats which will definitely fit, two by Corbeau and one by Ridgard so will limit comments to these seats.

The Corbeau GTA Clubman seat falls somewhere between cheaper sports seats and very expensive race seats, which makes it a good choice and excellent value for money. It works well with a full harness and fits into the car okay, albeit quite snugly.

A much better seat, but presently retailing at over three times the price of the Clubman, is the Sprint competition seat also from Corbeau. The Sprint is also available in Kevlar and carbon fibre, with fire-resistant Nomex trimming options.

The Corbeau Sprint seat has the all-important harness slots.

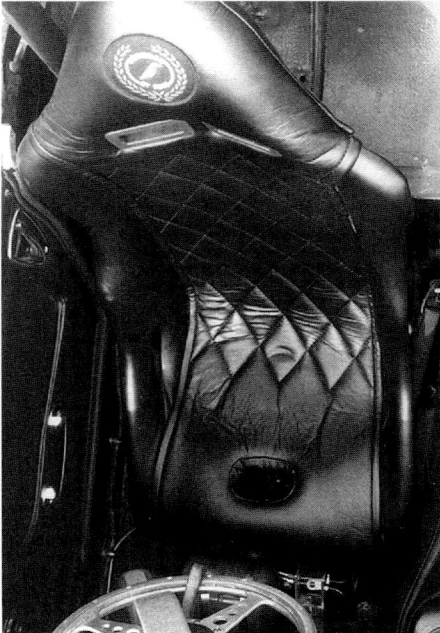

Corbeau Sprint seat re-trimmed in black leather.

This is an amazing seat which fits extremely well in the car. The all-important

shoulder support area of the seat is sufficiently narrower than other seats to clear the roll bar, but it does fit very tightly to the B-post.

Ridgard period seats are designed by Autostorica, and look the business for any Midget/Sprite. They're suitable for competition and are available with full harness slots, if required. The seats are available in a range of different coloured trim, including leather and contrasting piping.

Period-style seat from Ridgard: this is the 'Rally' which can be supplied either in vinyl or leather and with or without a headrest.

Aftermarket sports and racing seats usually use a universal or dedicated seat runner kit. Although fitting instructions are generally clear, you might like to consider the following points. It's often possible to reduce the overall height of the installation by shortening some of the seat brackets without weakening the fitting. Alternatively, you may be able to fit the seat runners direct to the floor pan, though this can be a tricky job. The best way to accomplish a direct fitting is to mark out the floor with a grid of masking tape, and then to mark bolt hole positions. Cross-

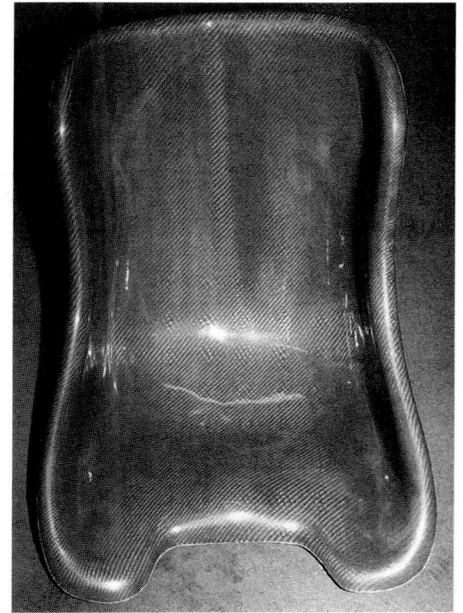

Carbon fibre kart seat from Carr Reinforcements - strictly for the racer.

check the installation before drilling the holes (be very careful not to drill through the petrol and brake pipes which run along a rail underneath the floor pan).

Corbeau also has a alloy subframe kit for the Sprint which is lighter but not in-car adjustable for back and forward movement.

If you are using any of the Corbeau seats and its bright trim is a little out of character for a classic race or rally car, this need not be a problem. Corbeau can re-trim your existing seat (or a brand new seat) in black leather. Leather is extremely hard wearing, smells great and adds period character. Corbeau has a US outlet which stocks the same range as the UK company.

Earlier, it was mentioned that the Corbeau Sprint seat is available in Kevlar and carbon fibre but this is likely to be prohibitively expensive for the majority of club racers. An alternative way to save weight is to use a go-kart seat. Although this kind of seat is not the most comfortable to use, it's acceptable for short, circuit races, sprints and hillclimbs. The ultimate lightweight seat is a carbon fibre kart seat and this is just what turbocharged Frogeye racer Simon Page uses in his racing car. The seat fitted both him and the car with ease, though the seat did require modification to allow the use of a full race harness. This type of seat is available from Carr Reinforcements Ltd.

FULL RACING HARNESS

An important consideration when it comes to competition preparation of your Midget/Sprite will be the racing harness.

There is a wide choice of harnesses and, generally, they differ by the number of straps and fixing points: the greater number of fixing points the more the load is spread through the car, representing greater driver security.

The Willans Club 6-point fixing, 6-strap harness and the similar 6-point harness from Luke are of good quality and both companies have sound reputations. The main visual difference between the Willans and Luke harnesses, is the option

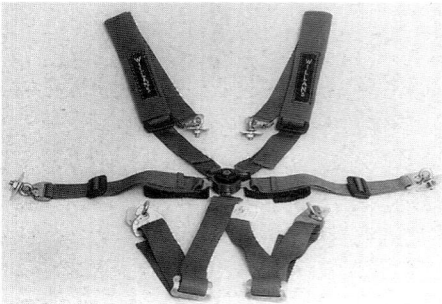
6-mounting point harness is superior to 3-point harnesses. This one is by Willans.

Luke harness has classic aero-type release box.

Old, standard seat belt mounting points can be used for some harness fittings.

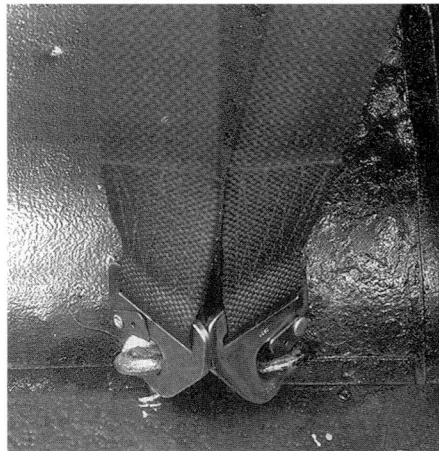
These rear bulkhead mounting points are close together: it would be better if they were spaced further apart to spread the load.

of a period (aircraft-type) release box on the Luke harness in keeping with the character of a classic car.

The Willans harness is normally supplied with the release box on the right-hand lap strap; if you're left-handed, you'll find this less than ideal. However, not only can Willans fit the release box on the left-hand lap strap, or even the crutch strap, it's also possible for the release box to be fitted rotated by 180 degrees - perfect for left-handers! (The same applies to the

Willans mounting ring and reinforcing plate with belt clipped in and locked.

Luke harness). If you have an existing, standard option harness it can be modified by Willans or Luke as appropriate.

To fit a multi-point harness it's necessary to decide where the mounting points will be. On the Midget/Sprite, these will invariably be on the floor and the rear bulkhead. A point to watch about the drilling of the mounts on the floor is to ensure you don't drill through the rigid brake line or rigid fuel line which will be very close to the ideal mounting points. Likewise, watch out for the rear fuel lines adjacent to the fuel pump when drilling the rear bulkhead. All of these areas (assuming your car is structurally sound) should be sufficiently strong to be load-bearing in the event of an accident, but you may wish to add (by welding) small reinforcing plates for extra security. Make sure you follow the seatbelt manufacturer's instructions, in particular with regard to positioning the eye bolts and also care and use of the harness.

BATTERY MASTER ISOLATOR SWITCH

Another standard safety requirement for

most types of motorsport is the fitting of a battery master isolator switch.

It's important to choose the right switch for your car. A standard battery master isolator switch will isolate only the battery so, if your car has either an alternator or dynamo charging circuit, the engine may well continue to run. If your car does have a charging circuit, you'll need an FIA-approved battery master switch which has spade end type connector terminals in addition to the main power feeds. If you have no charging circuit the standard switch will suffice.

Having purchased the appropriate switch for your requirements, it's relatively straightforward to fit. A good site on A-series-engined cars is the top edge of the right-hand front wing, in the corner, adjacent to the scuttle panel (assuming you have retained the battery in the standard position). For 1500 Midgets, use the left-hand front wing. These switch locations will not obscure your view and do not require any wiring to pass through the bulkhead. However, should your car be fitted with a one-piece front end, then it's probably best to mount the switch on the bulkhead.

To fit the switch, drill the holes and bolt the switch in. Wiring the switch will require removal of the battery live lead (first disconnect the earth lead). Two new leads (heavy-duty battery type) are substituted for the single live lead and these can be made up by yourself or an auto electrical specialist. One lead runs from the battery live terminal to the switch and the other runs from the switch to the main battery feed point (usually the solenoid). To keep the leads tidy, use P-clips or large wiring clips. If you can get hold of suitably-sized connection covering boots, use them (solenoid type rubber boots fit). Check that the switch leads do not foul the body once fitted. Refit the earth lead last of all.

An FIA-approved switch needs extra wiring to function correctly. Of the four extra connectors on this type of switch, two are for the ignition and two for the charging circuit (of particular importance for alternator-equipped cars, to protect the

Standard (left) and alternator-type battery master isolator switches.

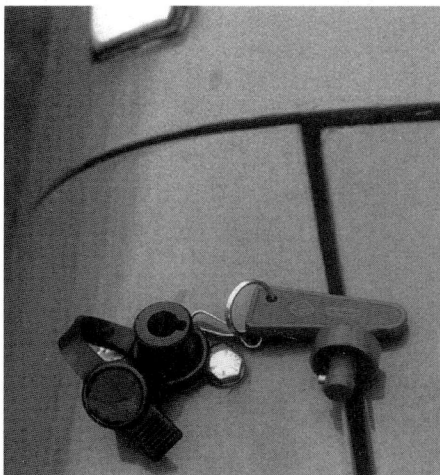

Switch shown with key alongside. The top edge of the front wing is an ideal mounting place.

regulator from damage from a high voltage surge if the switch is opened whilst the engine is running). Incidentally, if your car's alternator is damaged in this fashion, the regulator (Lucas part number UCB100 for 16,17 and 18 ACR units) can be replaced for a fraction of the cost of an exchange or new alternator.

The Autolec switch contains fitting instructions and a simple wiring diagram which needs clarification (Autolec informs me that the diagram will be revised in due course). The alternator feeds to the solenoid do not all need to be wired through the "W" terminal; what is required is that a single wire must run from the solenoid to the terminal. The other "W" terminal is wired to earth via the resistor supplied with the switch. You can solder wires to each end of the resistor and then earth one, connecting the other to the

switch. The ignition switch to coil wire can be cut at a convenient point so it is wired through the "Z" terminals of the switch.

FLAMEPROOFING BULKHEADS

For any form of motorsport, it's more than likely that all the bulkheads on your car will have to be flameproofed.

The main area for attention is the cardboard rear divider between the boot and area behind the seats. Removing the cardboard is easy as it is held in place by screws only: try and remove it in one piece so it can be used to make a template for a replacement made of aluminium or steel. This can be difficult to fabricate yourself without special tools so I suggest that a local bodyshop do the job for you. Fixing can be by self-tapping screws or rivets but, remember, there must be no gaps.

Occasionally overlooked are the cardboard panels behind the doors that seal the inner wing space from the passenger compartment. These panels should be replaced by light gauge sheet

Carpet and trim removal will reveal rear inner wing access apertures which should be plated over for competition.

Heater ducting to car neatly blocked off on Geoff Hale's Sprite.

steel or aluminium (either a single plate or three separate plates). The passenger side is easy to fit using self-tapping screws or rivets. The driver's side is more difficult as the wiring loom for all the rear lights passes through the lower panel. You can drill an appropriately-sized hole in this panel and, after fitting a grommet, thread the loom through it.

The heater air outlet from inside the engine bay needs to be blanked off, too. Once more, a suitable plate can be made in light gauge steel or aluminium sheet. Of course, once it's fitted you won't be able to use the heater.

HAND-HELD FIRE EXTINGUISHERS

A good budget choice of extinguisher in this category is a simple 2.25 litre hand-held unit with mounting bracket (such as those available in the Lifeline or SPA Design ranges). Fitting is quite simple and there are only a couple of points to consider when choosing the mounting point. Keeping the centre of gravity low will mean a floor mounting. Ease of access for the driver or passenger in an emergency is also something to consider and, finally, the unit needs to go where it

will get neither too hot nor too cold. A floor mounting on the passenger side of the car, where the extinguisher is unobtrusive yet ideally placed in an emergency, is ideal. When drilling holes for the bracket, watch out for brake and petrol lines and the exhaust pipe.

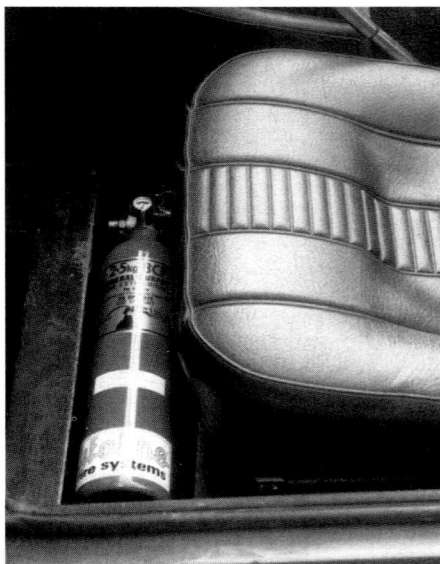

Fire extinguisher neatly mounted in author's car.

PLUMBED-IN FIRE EXTINGUISHERS

Plumbed-in systems generally used to use a 5.5lb (2.5kg) or larger capacity unit with BCF/Halon extinguishing agent. However, with the phase-out of BCF/Halon on environmental grounds, both Lifeline and SPA Design have developed water-based foaming agent which is lighter in weight than the old BCF/Halon equivalents. 2.25 litres of the new agent is equivalent 5kg of the old.

Tanks are made in steel (halon only), aluminium or carbon fibre so, when finances permit, carbon fibre or aluminium are obviously the better choices since they are lighter materials. As with a hand-held unit, the best site is on the floor (in order to retain as low a centre of gravity as possible). The ideal site is the passenger-side floorpan, assuming no seat is fitted (even if one is, there may just be room for a small unit in front of the seat). Mark, then drill the holes (watch out for brake and fuel lines and the exhaust system) that will be used to fasten the bracket to the car.

Lifeline units will operate in any position - even upside down - so do not

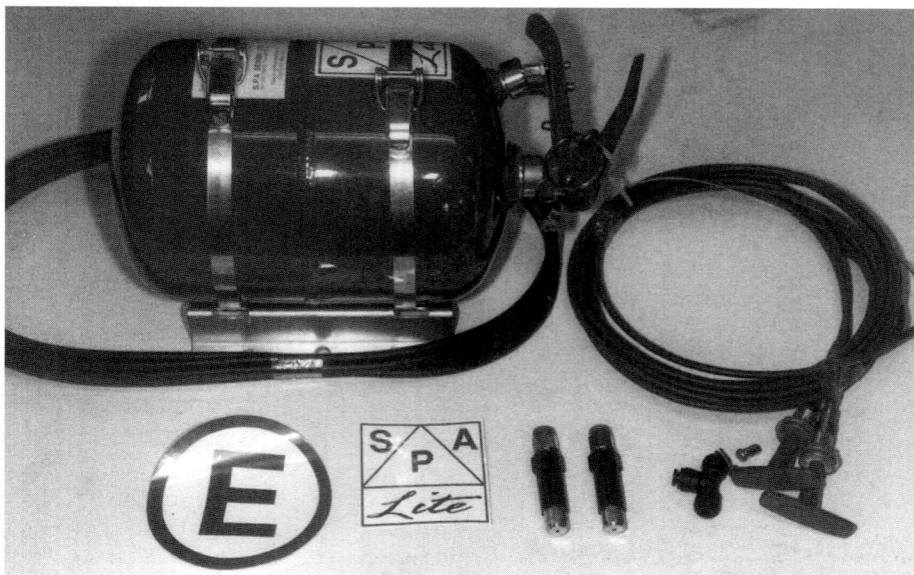

SPA fire extinguisher using non-Halon extinguishing agent. Environmentally friendly, and lighter, too.

Lifeline aluminium bottle plumbed-in fire extinguisher kit.

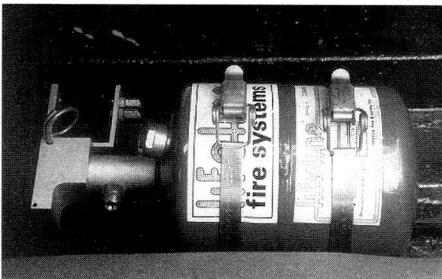

Lifeline 2.5kg (5lb) Halon bottle installed between driver's seat and body crossmember.

Close-up of Lifeline cable release on extinguisher head. Note that the safety pin is installed.

be unduly concerned on the orientation of the cylinder. Note, though, that this may not be the case with other manufacturer's units.

Interior pull handle for plumbed-in fire extingusher is mounted in dash.

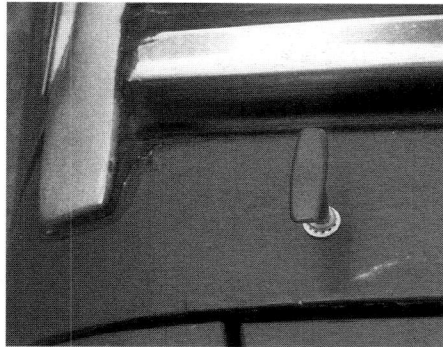

Exterior pull handle for plumbed-in fire extinguisher mounted on front scuttle panel.

Sites for the discharge nozzle(s) also need to be found. On the Midget/Sprite it's suggested that a nozzle be placed adjacent to the carburettor(s) and manifolds. It's possible with most kits to have a second nozzle. Usually the second nozzle discharges into the cockpit, though it could be positioned to discharge at a second point in the engine bay. Where appropriate, the second nozzle is fitted using the three-way 90 degree outlet supplied with the kit. Since the piping for the engine bay discharge nozzle is almost certain to pass through the engine bulkhead, a special fitting is contained in the kit for this purpose.

Plumbed-in systems can be either cable- or electrically-operated. As electrically-operated kits are quite a bit more expensive, they're an unlikely purchase for the club racer. Important instructions on cable runs, etc, will come with the kit. The system will have two cable releases: one, intended for driver operation, is mounted

inside the car and the second, for marshalls/rescue crews, is mounted outside the car. The cables need free movement, so check cable operation before connecting the cables to the extinguisher head. The cabling run needs to be as smooth as possible, so avoid any tight turns.

On some - but not all - extinguishers the head can be removed from the extinguisher, providing an opportunity to test the cable action.

As an illustration of how valuable these units are, the aluminium 5kg (11lb) Lifeline halon unit fitted to Simon Page's turbo Frogeye (featured often in this book) was used in anger during practice at a race meeting at Silverstone saving both driver and car. Damage was minimal and Simon raced the car that same afternoon from second place on the grid.

LAMINATED WINDSCREEN

It is a competition requirement that the car has a laminated windscreen instead of the normal toughened screen (which crazes when it breaks). Although the Midget/Sprite was originally fitted with either toughened or laminated glass, only laminated replacements are now generally available (for post-1964 cars) under Triplex part number AHA8008 B01602.

Fitting a replacement screen is not a job easily done yourself, even with experience of windscreen fitting on other cars. This is not so much because the frame assembly must be removed from the car (a job made easier by first removing the dash, but possible without), but because the frame has to be taken apart and the screws which hold it are often seized (new frames are not available). Help is at hand, though, as Peter May Engineering Ltd provides an exchange fitting service, although any competent windscreen specialist should be able to cope, assuming you have first removed the frame.

ROLL BAR

The Aley Aerodynamic roll bar is a

Two different types of roll-over bar, both of which offer similar strength. The diagonal brace of the right-hand roll-over bar is removable for non-motorsport use.

In the USA Tom Colby's Speedwell Engineering manufactures various designs of roll bar. Proof of the strength of these bars was a major motor racing accident at Willow Springs racing circuit when a racing Frogeye (Bugeye) landed heavily on its roll bar, then rolled a second time - ripping the front suspenson clean off the car. The roll bar and all the welded joints were intact after the accident.

SHELL PREPARATION

It's a good idea to strengthen the bodyshell by welding the seams. It may be that you even have a brand new Heritage shell to race prepare and, if this is the case and you're in the UK, Safety Devices can completely race prepare the shell on your behalf and to your specification. For instance, this company can weld in the roll cage and flare the arches, as well as seam welding the body for extra strength. All of the work that can be done on a new shell can also be done on an old one, but is made much more difficult by having to remove sealer, underseal, old paint, trim, *etc*, before serious work can start.

popular choice amongst Midget/Sprite owners and can be upgraded to an FIA race-approved bar (with removable diagonal brace) with no problems whatsoever. Aley Bars are made by Safety Devices Ltd., in England.

There is a large choice of roll bars for the Midget/Sprite. Typically, they are made from cold drawn steel (CDS). A lighter, but more expensive, alternative is to have the bar fabricated from aviation specification T45 material, which is both stronger and lighter than CDS. In the UK, Safety Devices can help you with this.

Which type of bar you purchase will depend on how you use your car. Roll bars for road use can simply be bolted in, but FIA-approved bars need welded plates to make them race legal.

In the UK, Safety Devices can fit your roll bar for you. Fitting times are in the region of an hour for the simpler, non-FIA roll bars and about half a day for welded-in plates and FIA-approved bars.

1967 Speedwell Sprite owned by Tom Colby. Note headrest pad incorporated in roll bar.
(Courtesy Tom Colby, Speedwell Engineering).

Left: Another very neat Speedwell Engineering roll bar. *(Courtesy Tom Colby, Speedwell Engineering).*

Chapter 14
Dynamometer Testing

INTRODUCTION

There are two types of dynamometer, or "dyno:" the engine dyno and the chassis dyno. The first is an engine test-bed dyno where the engine is run independently of the car and power readings are taken direct from the flywheel. The second type is where the engine remains in the car and power readings are taken from the rear wheels, via a pair of rollers. Because chassis dyno runs are taken from rollers, this type of dyno is often known as "the rollers" or "rolling road." Generally speaking, engine dynos are not available to the public, so this chapter confines itself to chassis dynos.

WHAT THE DYNO CAN DO

The dyno is the best place to discover certain things about your car's engine, such as real brake horsepower and torque output figures as measured at the rear wheels. You can also calibrate the speedometer and tachometer whilst, at the same time, learning what exhaust

A typical dyno station.

emissions the engine is producing.

Primarily, though, you'll be running your car on the dyno to tune the engine for maximum horsepower by obtaining optimum ignition and carburation settings under real-world running conditions. A point to note is that power figures obtained from the rear wheels cannot readily

be converted to flywheel figures. The other thing to remember about the horsepower figures you record is that they cannot readily be compared to power figures from other dynos. Finally, even consistent use of the same dyno can produce variations in horsepower figures. Therefore, the dyno should not be used to produce a set of figures to boast about in the paddock or bar, but rather as an instrument with which to get the very best result from your engine on the day you do your dyno runs.

A good dyno operator, such as Peter Baldwin, can tell you is if there is something wrong with the engine. The author took his tuned Sprite, complete with new unleaded fuel cylinder head, to Peter. On the first full power run the engine sounded a bit rough at 6000rpm, though this had not been noticed on the open road. Peter said it sounded "awful" and that it was probably the result of valve bounce. We did a second full power run while Peter listened at the carburettor intake in order to confirm his diagnosis, which he did. So, although the ignition and carburation settings were a good guestimate on the author's part, any use of revs much over

As well as providing information on engine horsepower and torque output, the dyno can be used to calibrate speedometers, too. Here, the author's car is doing a stationary 86/87mph.

Peter Baldwin (left) discusses engine specification with the author.

6000 would have wrecked an expensive engine. Oselli, who built this engine, have not yet commented.

A CHASSIS DYNO SESSION

The first thing to do before a dyno run is choose which of the many dyno operators to use. This choice may well be limited by geography but, whatever your choice, try and use the same dyno for future tests. Make sure the dyno operator is familiar with the Midget/Sprite and whatever type of fuel system you are running on your engine. Furthermore, if applicable, check that the operator has a suitable stock of needles or jets for the carburettor/s on your car.

The author has used the dyno at Marshall of Cambridge, England, for some years now and can vouch for the expertise of Peter Baldwin who operates it. In fact, a lot of Peter's clients are serious club racers who have achieved good results on the track. Peter is very familiar with the A-series engine, as well as both SU and Weber carburettors. The author has

always come away from this dyno with more horsepower and a smoother-running engine.

When you do a dyno run, always take notes to keep a record of the adjustments made during the session, and consequential readouts; these are invaluable for future reference.

Before the dyno run begins, tell the operator what changes you've made to your car's engine, ignition and fuel system. Discuss any problems experienced with the engine and what you're hoping to achieve from the session. This helps the operator know what to look for, and how to best spend the time.

The first job is to drive the car onto the rollers and get the wheels nice and central. The operator will chock the front wheels and then proceed to attach his instruments, which will usually include an exhaust gas analyser and various wires to the ignition system.

From this point on, in the author's experience, this is how a typical dyno test run will proceed. The engine is started ready for the first part of the session (a run in neutral just to check the ignition system is working okay). If, for instance, the engine has a bad lead or arcing distributor cap, it would show up on the tuning equipment. The first rolling run is usually done at a low engine speed simply to establish a good ignition setting. This is done by swinging the distributor while you're 'driving' the car. Next, the engine will be speeded up to the point where all of the distributor advance is used up (usually around 3500rpm for A-series engines). At this point the operator will be able to tell you what the static ignition setting will be. This setting will be unique to your car's engine and set-up. If you are really serious about ignition timing, it's possible to plot optimum readings right through the rev range. This information can then be used to adjust the distributor's ignition advance curve to the ideal for your car's engine.

Following on from the slow engine speed run the next one will see how the engine performs up through the rev range;

the operator will be watching the CO gas analysis to see if the mixture is too rich or too weak at any point in the rev range. The operator will want to know how the car is used as this will affect the final mixture strength settings. The author's car, for instance, has always run close to a full race mixture strength at full throttle, even though the car is only used on the road.

Once the carburation settings are correct, a full power run can be taken, which will give you power outputs as read at the rear wheels. You may even wish to take power readings at 500rpm increments to plot a power curve for your own use. Whatever you decide, the more information about your engine you can get, the better.

While you've been doing all your dyno runs, hopefully, you'll have kept a close eye on water and oil temperatures. Cars usually run hotter on the dyno,

despite the presence of a large fan in the dyno bay to cool your engine in the absence of the normal airflow.

When you've finished your dyno runs, the operator will often ask you to drive your car on the road for a short test to ascertain that everything feels okay. This is usually an interesting experience as the car will nearly always feel a lot smoother and more powerful.

The bill for the use of the dyno is usually based on an hourly rate plus the cost of jets and/or needles. The price is usually fairly modest, given the gains in horsepower you will have made and the data you have gathered.

Any further development of your car's engine should always be followed by another visit to the dyno. Where finances prohibit this for any length of time, keep a record of the changes you've made and what affects they had.

Chapter 15
Bodywork

INTRODUCTION

Before you begin any bodywork modifications, your car MUST be structurally sound. Help is at hand in the form of brand new BL Heritage bodyshells for the Midget/Sprite at a very reasonable price. However, Heritage shells are not available for cars with a chassis number prefixed earlier than GAN 4 or HAN 9 although, as this book went to press, Classic Automobiles of Rutland, England were producing new Frogeye shells which have been adapted from the Heritage unit.

WHEELARCHES

Front wheelarches will probably need to be flared outward to accommodate your chosen width of wheel and tyre combination at full lock and full bump. The following method will produce a neat flared arch.

Cut the existing arch with tin snips at regular intervals, each cut being at 90 degrees to its section of wheelarch, then peel out the sections to the amount of flare you require. The segments should be dressed with a dolly and hammer to the required shape. Carefully tack weld the segments and then dress again with the hammer and dolly as required. Finish off the arch using a light skim of body filler or lead loading.

Rear wheelarches on the Midget/Sprite can be square or round depending on the age of the car. Round arches can be flared using the method already described. Square arches, on the other hand, are quite hard to flair and it is

Work in progress at Carcraft on flaring Sprite front wings. *(Courtesy Carcraft).*

Flared front wheelarch on the author's Sprite is subtle and neat - work carried out by Carcraft.

convert your car's body to round wheelarches; this can be done by using new wings or repair panels.

USA-based Sports and Classics manufactures rear arches in good quality fibreglass for 1962 and later models.

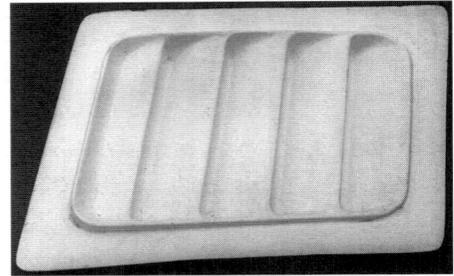
Typical fibreglass louvred vent panel.

easiest to raise the arch height so as to retain the original look but allow the use of a wider tyre/wheel combination. This can be a difficult job and would, ideally, be undertaken during renovation of inner and outer wings. An easier solution is to

Joe Huffaker's F Production racing Sprite. Note front air dam, headlamp air duct and flared arches: just a few features of this impressive race car. *(Courtesy James Martin Photo)*.

BONNET LOUVRES (VENTS)

Engine bay temperatures can become a problem in hot weather. The easiest solution is to let some of the hot air out through the bonnet and this can be done by the use of louvres. Louvres have been used on production cars as diverse as the Jaguar E-type and the Ford Sierra Cosworth. It's possible to go for the traditional punched metal look or the modern fibreglass look.

Fibreglass accessory-type louvres come in many shapes and sizes and can be fitted into appropriately-sized holes cut in the bonnet. Cutting holes in steel bonnets may be tricky to accomplish, but will be relatively easy with the fibreglass bonnets and one piece fronts commonly fitted to Frogeye Sprites and racing Midgets/Sprites. Louvres are fixed in position by rivets/self-tapping screws and

Louvred bonnet prior to painting and fitting.

Louvred bonnet fitted on car - louvres are sited in exhaust manifold area.

Cutaway front wing allows hot air to exit and may reduce drag, too.

Louvred vent panels improve airflow under one-piece bonnet.

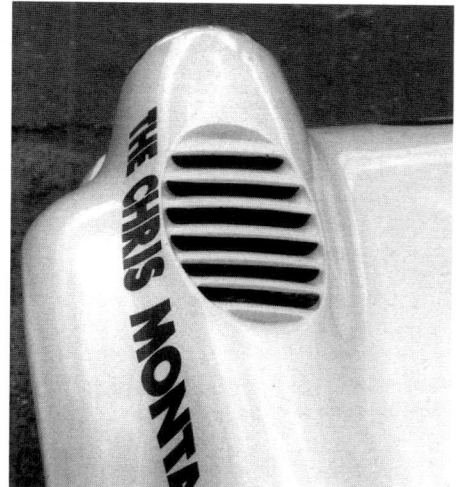

Louvre vent in front wing top is designed to extract air from wheel bay (theory is to reduce air pressure and thereby improve downforce).

Exit for hot air that has passed through the radiator on a racing Midget.

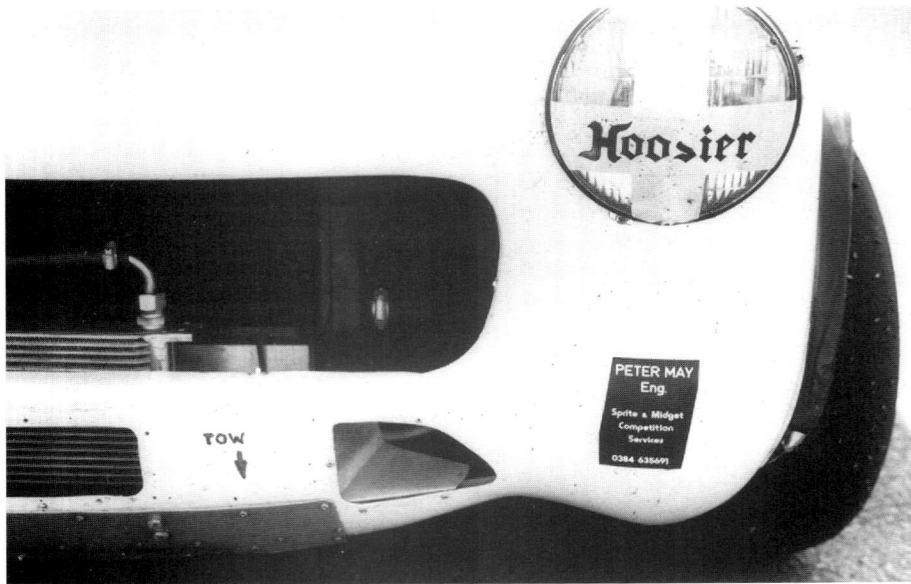

Ex-Peter May race car has tidy air ducting for front brakes, also cut-out in front panel to increase air flow to low mounted (minimal obstruction of radiator) oil cooler.

Necessary relief of inner wing to give clearance for a DCOE carburettor using 39mm full radius rams and K&N filter.

then 'glassed-in' to finish.

Real louvres can be added to steel or aluminium bonnets by a special tool in a machine press. One company that can do this is Kool Louvre (run by Robert Pocock) in Aylesbury, England. Kool Louvre can punch louvres from 1 inch up to 5 inches (25-381mm) in width and there is no limit to how many - other than the strength of the bonnet! You tell Robert the positioning, number and size of louvres and he does the rest. The author took a new Ron Hopkinson-supplied bonnet down to his workshop to have louvres punched. He settled on a double row of louvres over the exhaust manifold towards the rear edge of the bonnet (avoid positioning the louvres within 12 to 18 inches/300 to 460mm of the windscreen) which allows the hot air to escape really well. For your own car you may, of course, wish to experiment and find a good low pressure area in the airstream or maybe even try the middle of the bonnet.

Circuit race cars, hillclimbers and sprinters may require a much bigger vented area, or areas, than a road car. Some interesting racer's solutions to venting engine heat are pictured in this chapter.

BONNET PROP

For any non-Frogeye Sprite or non-1500 Midget the bonnet prop consists of a single rod that hooks into a slot on the inner wing; it's prone to rattling loose with the result that the bonnet lands on your head as you work on the engine! The 1500 Midgets get around this problem by using a later design of stay which extends to support the bonnet when it's open and is closed by releasing a bonnet catch. This catch can be fitted to any earlier car by purchasing the later-type stay with the two round sleeves that fit at each end.

CARBURETTOR, AIR FILTER AND RAM TUBE CLEARANCE

Long(ish) rams are used for maximum engine flexibility as well as horsepower. Unfortunately, this means that a fairly large air filter housing is needed to accommodate the rams and this, in turn, means that the inner wing will need to be relieved to make space.

Aside from the relief of the inner wing, a very large air filter may also require some relief at the bonnet edge.

THROTTLE LINKAGE CLEARANCE (WEBER DCOE)

If you have fitted a Weber carburettor and a Janspeed throttle linkage, you may be in need of this short paragraph. You have two choices; either you modify the linkage or you'll need a bonnet bulge to prevent the linkage jamming against the underside of the bonnet. One way of achieving the latter option is to cut a hole and fibreglass-in a suitable pre-made bulge of your own design (which can have been formed on any suitable ball or curved object). Another option is to use the TR4-style power bulge from Fibresports. By far the neatest option, though, is to take the bonnet to your local bodyshop, where the job can be carried out professionally by an experienced panel beater.

TUBULAR EXHAUST MANIFOLD CLEARANCE

If you're fitting a tubular exhaust manifold, you may (will with the Janspeed 3-into-1 manifold) find that some extra clearance is required at the 90 degree bend and the bottom of the pipes. Cut back the metal as necessary to clear the downpipe and then, if required, weld the seam.

FROGEYE BONNET COVER (BRA)

Available only from Barrett Enterprises in

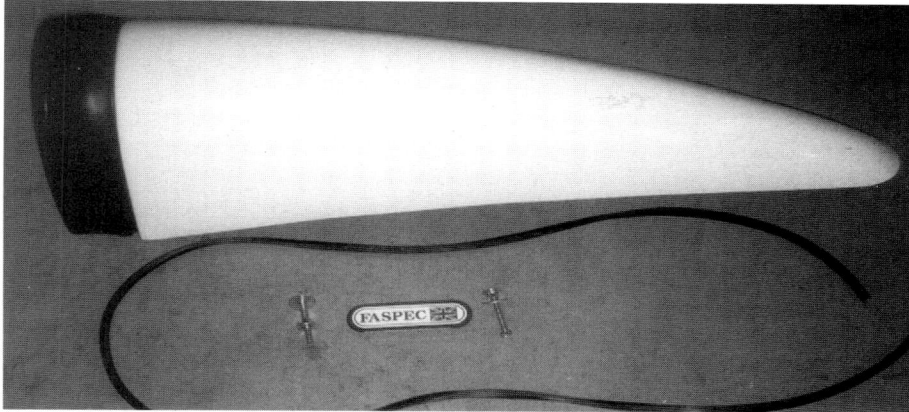

Side view of the Faspec Frogeye headrest fairing kit complete with rubber seal and fasteners.

your average Formula One car bodywork but, nonetheless, the weight saving was considerable and Simon was well pleased with the result.

ALUMINIUM AND FIBREGLASS PANELS

Aluminium is much used by racers, but there is no reason why you can't have a lightweight road car. Motobuild lists a whole range of aluminium panels in its catalogue which includes front nose skin (between wings), boot lid skin, door pillar

the USA, this is a front end cover for the Frogeye Sprite which is made of the highest grade fleece material with felt binding. Ideal to protect the front end of the car from stone damage and even winter rock salt. Each cover comes with comprehensive fitting instructions and a storage bag.

FROGEYE HEADREST FAIRING KIT

Available from Faspec in the USA is a really neat-looking headrest fairing. The kit consists of a fibreglass fairing section, trimmed headrest section, a rubber sealing strip and fastening bolts.

Fitting requires two holes to be drilled in the car body. The fairing will require painting to match the car colour. The headrest is black with black piping but it would be quite easy to have it re-trimmed to match any existing colour scheme. The kit is designed for left-hand drive cars but can be modified to fit the right-hand drive cars. Bearing this in mind, there's no reason why you couldn't have a headrest fairing on each side of the car.

CARBON FIBRE BODYWORK

Although the use of carbon fibre may sound a bit exotic for a club racer, several racers do use carbon fibre doors on their cars. The advantage of carbon fibre is its

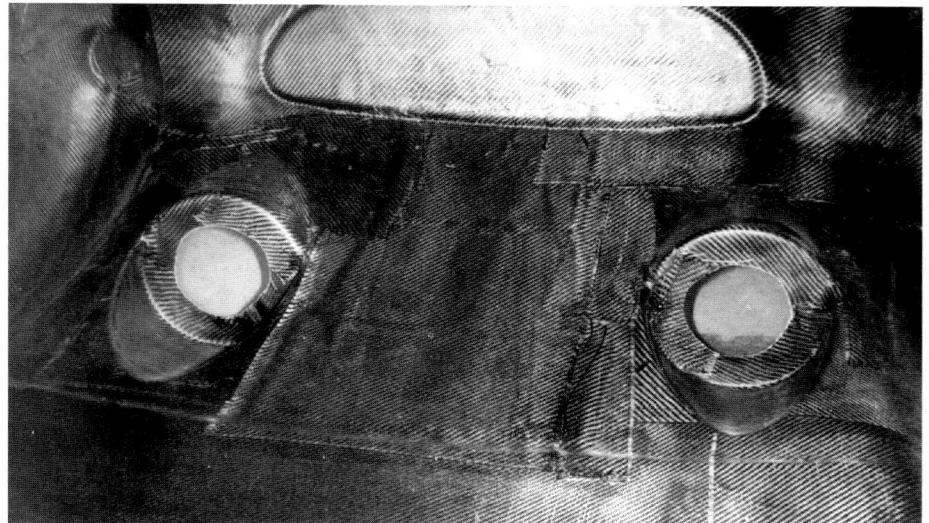

One-piece Frogeye carbon fibre front by Mini Classic Cars/Carr Reinforcements. Substantially lighter than the fibreglass equivalent.

lightness combined with great strength.

Any panel currently manufactured in fibreglass could be produced in carbon fibre if you have access to the relevant moulds. A company that can supply the necessary raw material, and sometimes produce the finished product too, is Carr Reinforcements Limited in Cheshire, England. Carr Reinforcements referred the author to Mini Classic Cars who manufactured a one-piece front for Simon Page's racing Frogeye. Because of the condition of the mould Simon supplied they found it necessary to use a gel coat before laying up the carbon fibre, which made it marginally heavier and less exotic than

outer skin. For some applications, like racing, you may wish to fabricate the parts yourself. One person who's done just that is MGCC competitor Robin Griffiths, whose car has a lovely one-piece aluminium front. It really is one-piece, being fabricated from a single sheet of aluminium.

A fibreglass one-piece front will give new meaning to the term "engine accessibility" and can alter the appearance of your car, too. Motobuild can supply one-piece front and rear ends as well as all the necessary fittings. In the USA, Speedwell Engineering has a good range of one-piece fronts.

SPEEDPRO SERIES

Above: One radical approach to lightweight doors. The car is Dave Grove's Sprite.

Front spoilers are available from Motobuild in the UK or Sports and Classics in the USA.

FULL BODY CONVERSIONS

Arkley

The Arkley conversion has recently become available again from Arkley Sportscars (which is not the original company). In the USA, the Arkley conversion is available from North American Arkley. The British Arkley conversion for the Midget/Sprite is called the Supersport and is available as a full kit or front and back panels only; the latter being very handy if either end gets accident damaged. The full conversion kit comprises the following parts: body kit, headlight units, front indicators, rear lights, rear indicators, 7in wheels and caps, 185 x 60'tyres, bonnet fasteners, flasher unit and sidelamp adapters. Arkley recommends that, for safety reasons, a structurally sound donor car is used for the conversion. Any model of Midget/Sprite can be used, the only difficulty being that the hood from early cars cannot be used with the kit. Of course, now that new bodyshells are available, you could always build an Arkley from scratch using a modified new shell.

Speedwell Sprite

A rarer conversion is the Speedwell. Before its demise, the original Speedwell company made numerous tuning parts for the Midget/Sprite, the most interesting of which were bodywork conversions and bonnets. Richard E. Rooks in the United States has researched the Speedwell story in some depth and can supply a copy of his findings. Speedwell conversions varied in appearance but I can say that, should you come across a car which looks like a

Speedwell aluminium air shroud for fibreglass front end. *(Courtesy Tom Colby, Speedwell Engineering).*

Arkley conversion from Arkley Sportscars. *(Courtesy Arkley Sportscars).*

122

Ferrari 250GT but has Sprite running gear, you've probably discovered a Speedwell Sprite. Richard's own car not only has the full body treatment (including a permanent hardtop), but also has a Fiat 1500cc engine. Richard is compiling a register of Speedwell Sprites and will be pleased to hear from owners of similarly modified cars.

Sebring Sprite

A product of 1950s and 1960s racing and rallying of the Sprite was a special car known as the Sebring Sprite, named after the circuit of that name in Florida, USA. Noted Sprite tuner of that period ,John Sprinzel, produced a rebodied Sebring car with completely restyled bodywork. However, only six cars with this original bodywork were made. A Sebring replica can now be built using parts manufactured by Archer's Garage in Birmingham. Three main body parts are required to make a Sebring replica, a Sebring bonnet, Sebring coupe roof and a tail section. To complete the conversion special doors, sidescreens

and Donald Healey speed equipment seats are also available from the same manufacturer.

Sebring Sprite one-piece front end from Archers Garage Ltd. (Courtesy Archers Garage Ltd).

Unlike the original conversion, Sebring replicas can be made in fibreglass, diolen or kevlar: all of which have the

Walker front end by Tom Colby's Speedwell Engineering: car is owned by Tim Walker. (Courtesy Tom Colby, Speedwell Engineering).

Left & below left: Roof and one-piece rear section for Sebring Sprite. *(Courtesy Archers Garage).*

FROGEYE CAR COMPANY

Originally this company were offering fibreglass kit-based conversions for rusty and tired Frogeye Sprites, but has now gone on to produce a brand new car.

If you can imagine a Frogeye Sprite with a fibreglass body, 1275cc engine, Ford gearbox and disc brakes as standard, you have a good idea of what the Frogeye Car Company is all about. Think brand new rather than donor car, because even the engine is brand new. It uses an A-Plus block with a special flywheel, all put together by Oselli Engineering. The front suspension is a twin wishbone set-up with coil over shocks linked with Metro brakes.

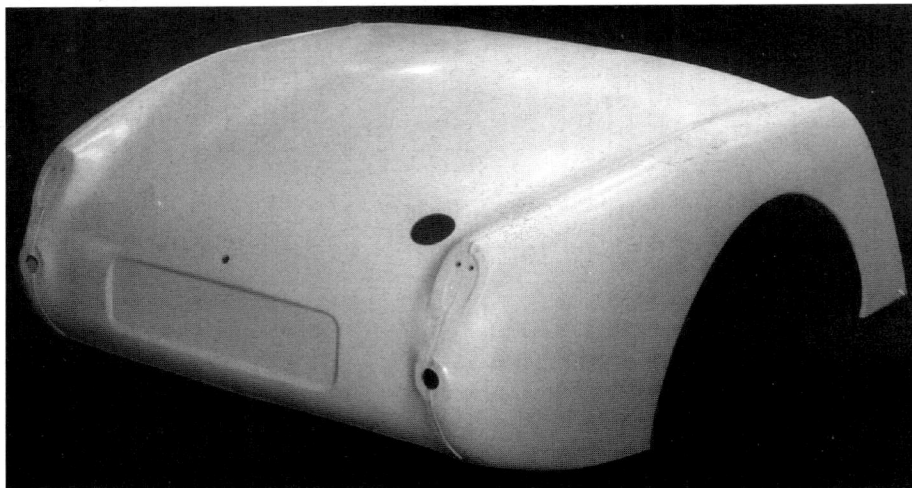

BONNETS FOR FROGEYE SPRITES

The demise of the original Speedwell Sprite is not the end of the story. Tom Colby's Speedwell Engineering in the USA took a mould from a 1959 Speedwell Roadster, chassis number 6, and now produces Speedwell Monza bonnets for Sprites. In the UK the original Williams

Below: One-piece front end on Graeme Adams' racing Class A Midget. Note ducting to radiator.

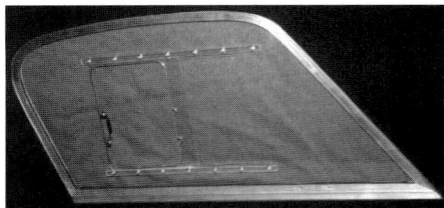

Side screens for Sebring Sprite. *(Courtesy Archers Garage).*

further option of being supplied with a fire-retarding resin. Archer's Garage can also supply brand new (not rebuilt) Sprite MkI bodyshells with the non-standard option of half-elliptic spring mountings. At least one complete kit has gone to the USA and will be assembled for a customer by Tom Colby's Speedwell Engineering.

Tom Colby, proprietor of Speedwell Engineering, USA, racing at Willow Springs International Raceway at 118mph. Note the Speedwell Monza front end. *(Courtesy Charles R Colby).*

Below: Dzus fastener panels for fibreglass one-piece front end. *(Courtesy Tom Colby, Speedwell Engineering).*

Simple fabricated bracket (utilising towing eye mounting captive nuts) acts as hinge for one-piece front.

and Pritchard mould was purchased by Fibresports Ltd. who can still make Monza bonnets, which are available only through Perfect Nostalgia Ltd.

Tom Colby's Speedwell Engineering has another neat bonnet with the head-

David Clarkson's supercharged Frogeye still wears its original one-piece front. If only they'd kept the idea for Midgets and later Sprites.

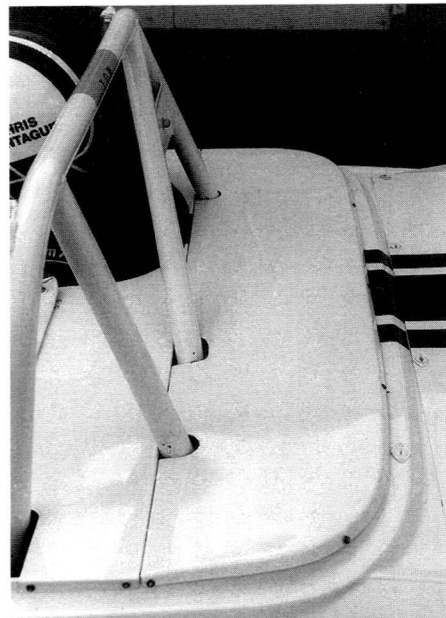

Panel 'tonneau' cover on Chris Montague's racing Class A Midget.

lights mounted outboard in the wings. This is called the Walker bonnet, and is obviously more aerodynamic than the standard bonnet. Fibresports has a front called the Sebring, although this is not the Sprinzel/Speedwell Sebring front which it can also make.

MIDGET/SPRITE ONE-PIECE FRONT ENDS

In the UK, Fibresports can supply a Midget/Sprite (post-Frogeye) front end, as can Tom Colby's Speedwell Engineering in the USA.

HARDTOPS

There are numerous hardtops available for the Midget/Sprite. However, the most radical is the Fibresports fastback hardtop which is almost a body conversion: it makes a Midget/Sprite look like an MGB GT. Fibresports also makes more conventional hardtops as well as having moulds for some other very neat and unusual tops.

LOCKS

The Midget/Sprite, unlike most modern

cars, uses a key for each lock on the car and, if a locking petrol cap is included, this means a bunch of five different keys. It is possible to change the locks and barrels on your car so that you have one, or for post 1970 cars two, keys for all the locks on your car. At the same time you can fit upgraded door locks which fit more securely by utilising a threaded body (held in the door by a large hexagonal securing nut) which prevents them from working loose as the originals do. These replacement locks are supplied with a large BL (British Leyland) key with a double-sided blade. This key-type was standard on Triumph models, and it is much more convenient to use

There are two ways to match-up the locks. The first and easiest is simply to order a full set of new locks which will include a replacement boot handle, two door "private" locks (LH & RH) and an ignition barrel (the latter is easily replaced into the existing Lucas ignition switch on pre-1971 models). Locking petrol caps were not specified as an original fitment, but there is an historic Wilmot Breedon model available from most dealers, which

Standard door lock assembly (left) and superior Triumph item.

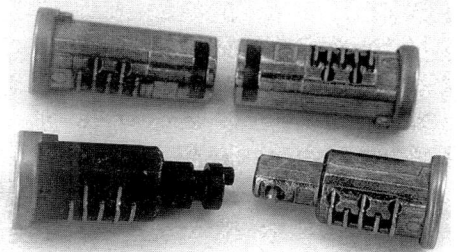
Lock barrels at top are for doors, barrel at bottom left is for boot lock and last barrel is for ignition switch.

Rarity! Fibresports fastback on Carl Bintcliffe's 1500 hillclimb Midget. Note wheelarch flares and front panel mounted spotlights. (Courtesy Carl Bintcliffe).

Wilmot Breedon locking petrol cap has period look and uses the same key as all the other locks.

can be matched to your car set, using the same "FS" keys

The second and more difficult method is to obtain a set of replacement lock barrels, and then to fit them within the existing lock units. This is a relatively easy task if your locks are in good condition but, if corrosion has developed, you may find barrels are difficult to remove and the chromium plating is pitted or the diecast housings badly oxidised.

On cars made after 1971, the ignition switch is incorporated into a steering column lock which uses a high-security key type which cannot be matched-up with the car's other locks.

Fitting lock barrels

The easiest lock barrel to fit is the ignition. Disconnect the battery earth lead. Unscrew the ignition lock securing ring on the front of the car dash panel and remove the ignition lock (preferably without disturbing the electrical connections). Insert the key in the lock, but do not turn it. Next find the small hole set in the body

of the lock and insert a scriber or similar pointed tool and press gently. While doing so, pull carefully on the key and withdraw the lock barrel. The new lock barrel can be fitted, again with the key inserted until it is all the way in, whereupon you will hear a click as it fastens in position. The key can now be withdrawn without the lock barrel falling out of the lock and the lock can be refitted to the dash and the battery earth lead reconnected.

To change the boot lock barrel, start by using a small chisel tap off the steel bellwasher from the lock shaft (you may need to file-off some of the nibbled material which holds this in place). Once this has been done, withdraw the handle and lock from the boot lock main body. Look for a small retaining pin in the base of the lock and tap this out with a scriber or similar tool. The lock barrel can now be removed and replaced with the new barrel.

Fitting new locks

Whether you are fitting old or new type door locks, you'll need to remove the original ones first. Strip the door trim and, using a screwdriver to release the retaining clip, remove the lock complete. If you are fitting the original type lock, simply push fit the lock into the door (ensuring the lock forks have interlocked with the door lock catch). Try unlocking the door a couple of times before refitting the trim.

If you're fitting new type locks (part number 9/01340), once the old lock has been removed you'll need to carefully enlarge the lock aperture in the door to allow for the extra size of the new lock. Once the lock fits, use a small paintbrush to touch in the bare metal to prevent rusting. These newer-style locks are generally used for MGB and TR6 models, both of which have a greater door thickness than Midgets/Sprites. Because of this, before fitting this lock to the Midget/Sprite, it's necessary to reverse the lock fork. This is easily achieved by removing the ring clip and letting the retaining pin drop out enabling the lock fork to be removed and reversed. Test fit the lock to see if it works okay.

If you find that the lock will not lock, it may be necessary to strip the lock and file the edges of the lock body to allow greater movement of the lock fork (when attached). Test fit the lock once more and if everything is working as it should screw the retaining nut onto the lock body. If you experience difficulty in tightening the nut, being unable to get a spanner on it, use a small chisel to gently tap the nut round until it is tight. All that remains now is to refit the door trim.

The boot lock is replaced in much the same way as the door locks. If you require full instructions, they can be found in the workshop manual.

Petrol cap lock

The petrol cap can only be matched to the rest of the locks if it is of the Wilmot Breedon type. The retaining screws on the lock need to be undone and the barrel removed and swopped for a matching barrel. Alternatively, you can easily tap-out the five brass tumblers and shuffle these to match-up to your key, filing if necessary: all the tumblers should lie flat inside the barrel when the key is inserted.

The easiest way to match this lock to the others on the car is, of course, to order a Wilmot Breeden locking cap as part of a complete lock set! This cap will not fit US "Federal" cars from about 1972 onwards.

Disabling interior locks

A modification that can be carried out while the door locks are being changed is to disable the interior door lock catches. One reason for disabling them is to prevent anyone gaining access to the car by removing the hood fastenings and putting a hand round to release the lock. Disabling only requires the locking lever on the escutcheon to be disconnected by removing the locking fork.

The disadvantage of this modification is that the passenger door can only be unlocked from the outside and by the key.

Chapter 16
Competition Cars

INTRODUCTION

Although you will most likely be modifying a road car, it's interesting to be aware of what it is possible to do with a competition Midget/Sprite. This short chapter looks at a few racing cars and a rally car that have achieved a good measure of competition success.

TURBOCHARGED FROGEYE SPRITE

Cars don't come much more unusual - or,

The power plant of Simon Page's Frogeye Sprite.

Auxiliary radiator once used on Simon Page's car (it was removed when single Serck aluminium alloy radiator fitted).

indeed, modified - than Simon Page's racing Frogeye Sprite and there are two items on his car that are of particular interest: the Toyota 5-speed gearbox conversion and the turbocharged engine. Simon initially used the Metro Turbo system 'off the shelf' although, at a later stage, he adapted the conversion to include a modified Montego intercooler. Simon uses temperature measuring stickers on the intercooler fittings to monitor effective temperatures, an idea usually associated with braking systems.

Superficially, the idea of

turbocharging a Midget/Sprite seems relatively straightforward. However, Simon's car demonstrates the true

Fabricated rear suspension and Ford Anglia back axle in Simon's much-modified Frogeye.

Drilled rear discs on Simon Page's Turbo Frogeye.

complexity of the system though, if you're determined, there's no reason why a road car cannot be adapted in the same way as Simon's racer.

Simon's installation is a good deal more sophisticated than the standard Metro system. The turbo installation obviates the need for an LCB-type manifold: the exhaust pipe from the turbo housing runs through a silencer to a short and tidy pipe which exits at the side of the car. The turbo system requires a non-standard fuel system which uses twin pumps (one low and one high pressure), one at each end of the car.

The Toyota gearbox is suited to the job of handling the output of the turbocharged engine in terms of both power and torque. The power is fed through an AP single plate sintered race clutch; the Toyota conversion bellhousing being relieved to accommodate this latter item.

It isn't just the engine and transmission that are unusual on this car, though, as both front and rear suspension is very well developed. Of particular interest was

Intercooler intake and outlet on Simon Page's Turbo Frogeye. Note temperature sensing strips.

the use of a Ford Anglia axle with a Quaife differential. The axle installation is one of a pair that was designed and built by Simon and a friend - Chris Norris. The axle uses Spax coil over shock absorbers and a Watts-linkage with anti-roll bar - all fully rose-jointed. The front suspension is a little more conventional with a modified Armstrong lever arm unit that has a second top suspension link. The front anti-roll bar links are rose-jointed. Each front hub carries a Ford Cortina brake disc (drilled) and calliper. All the brake linings are Ferodo.

RACING MIDGET I

Peter May's race car may be a lot less flamboyant than Simon's but, nevertheless, still has several interesting features. Being in the business of preparing race cars, engines and gearboxes is an advantage when running your own car and Peter's car is really well prepared. Although running a

Simon Page racing his much-modified Frogeye Sprite.

Peter May's (now sold) race car in the paddock at Silverstone. Note the carefully crafted NACA-type duct in the door and air outlet at the rear of the front wing.

One of two engine steady bars on Peter May's racing car.

conventionally tuned engine, the car has several interesting features worth mentioning.

On the rear suspension Peter has modified the Spax telescopic conversion kit so that the shock absorbers act in the vertical plane for optimum performance: Peter can sell you the appropriate modifications. Peter also sells Panhard-rod kits but on his own car he uses a specially fabricated item that is welded, rather than bolted, to the axle case.

In the engine bay the car uses two engine steady/tie rods which are rose-jointed. These rods prevent excessive engine movement.

The car uses a large disc brake conversion but, in common with most racers, does not use a servo.

The car's body is very tidy and the author particularly liked the cut-out and ducting to direct air onto the front brakes.

The car is now no longer owned by Peter, but its new owner is achieving good racing results with it.

RALLY MIDGET

Phil Bollen's rally Midget is a fairly standard car due to the class of rallying he

Phil Bollen's 1967 MkIII Midget which is used to compete in historic rallying.

Homemade, but nevertheless effective, is the sump guard fitted to Phil Bollen's rallying Midget.

competes in.

The engine modifications include K&N air filters, LCB manifold, RC40 exhaust silencer, alternator conversion and an oil cooler. Phil's car also has a Ron Hopkinson handling kit, racing seats, harnesses, a battery master isolator switch, Kenlowe fan and a sump guard. The rest of the car is pretty much standard.

The reason this car is featured is to illustrate that you don't need huge amounts of money and enormous technical expertise to have a lot of fun with your Midget/Sprite.

RACING MIDGET II

The author's initial interest in Eric's car was due to its big disc conversion. However, it's in more than just the braking department that this car is highly modified.

Eric has used a big disc conversion in conjunction with a 1500 Midget dual circuit brake master cylinder. The rear brakes use standard shoes.

The front suspension is particularly interesting, being to Eric's own design and using twin wishbones with coil over shock absorbers: a very neat arrangement. The

Note grille treatment and cut-down aeroshield-type windscreens of Eric Grundy's beautifully prepared racer. *(Courtesy Eric Grundy).*

Eric Grundy's Midget side-on; note sponsors include Manx Airlines since car is Isle of Man based. *(Courtesy Eric Grundy).*

Engine bay of Eric Grundy's car; note aluminium inner wings. *(Courtesy Eric Grundy).*

Neat fuel bag tank and Facet fuel pump in Eric Grundy's Midget. *(Courtesy Eric Grundy).*

Extension of the rear 'deck' of Eric Grundy's Midget goes well beyond the rear roll bar mountings. Other racers have adopted this approach, but none as tidily as this - not a rivet in sight. *(Courtesy Eric Grundy).*

only drawback to the conversion is that it's the only one in existence and Eric wouldn't be persuaded to make a few more! The spring rates are around 330lb (149.69kg) and, if that sounds a bit soft compared to the rates recommend in this book, you have to bear in mind that this car is not only pretty light (a third of the weight of a normal Midget), it's also used on road rallies rather than circuits.

Fuel is held by a very neat bag tank in the boot.

The boot has been reskinned in alloy, as have the doors. Elsewhere the body features race weight fibreglass and has been seam welded for extra rigidity.

The car was first run in the 1992 season, during which development bugs were ironed out. Eric's car can be seen in action on his native Isle of Man.

Appendix I
Specialists & Suppliers

Here's a list of suppliers and specialists, which should help you to track down the parts and services you require. Please note that inclusion in this list does not constitute a guarantee of quality or performance, though it is hoped you'll find all of these companies helpful and offering good service to Midget and Sprite owners.

You'll definitely find it useful to belong to an MG or Austin-Healey club. Such clubs are found all around the world; find your local club by consulting your country's motoring magazines.

SPECIALISTS & SUPPLIERS (INTERNATIONAL)

Abingdon Motors
192 Annerly Road
Brisbane
Australia

Advanced Products (K&N air filters)
Wilderspool Causeway
Warrington
Cheshire WA4 6QP
England
Tel: 01925 36950

Aldon Automotive Ltd
Breener Industrial Estate
Station Drive
Brierly Hill
West Midlands
DY5 3JZ
England
Tel: 01384 572553

Anglo Parts
Brusselsesteenwg 245
B-2800 Mechelen
Belgium
Tel: 15 42 37 83

Anglo Parts
3 Bis Rue des Ecoles
59 254 Chyvelde
France
Tel: 28 26 61 00

Anglo Parts
Storkstraat 3
NL-3905 KX Veenendaal
The Netherlands
Tel: 8385 51334

AP Automotive Products
Brakes Division
Tachbrook Road
Leamington Spa
Warwickshire CV31 3ER
England
Tel: 01926 470000

AP Lockheed
Rue De Chateau D'Eau
45410 Artenay-Chevilly
France
Tel: 38 741314

AP Lockheed
BP 219
95 614 Cergy Pontoise Cedex
France
Tel: 1 34 303430

AP Lockheed
Winterhauser Strasse 89
Postfach 3225
8700 Wurzburg 21
Germany
Tel: 931 614070

AP Lockheed
Corso Marconi 160
17014 Cairo Montenotte
Savona
Italy
Tel: 19 501327

AP Lockheed
CPO Box 956
Osaka 530-91
Japan
Tel: 6 386 6871

AP Lockheed
PO Box 136
5140 AC Waalwijk
The Netherlands
Tel: 4160 34917

AP Lockheed
Apartado 144
Eibar
Spain
Tel: 43170412

AP Racing
Wheler Road
Seven Stars Industrial Estate
Coventry CV3 4LB
England
Tel: 01203 639595

A.P.T.
561 Iowa Avenue (Building A)
Riverside
California 92507
USA
Tel: 909 686 0260

A.P.T. Concessionaires Ltd
(A.P.T parts in UK, esp studs,
etc.)
Unit One
Rutherfords Business Park
Marley Lane
B attle
East Sussex TN33 0RD
England
Tel: 01424 772202

AP (USA) Inc
4000 Pinnacle Court
Auburn Hills
Michigan 48326-1754
USA
Tel: 313 377 6999

Archers Garage Ltd (Sebring
parts)
65 Pope Street
Cnr Icknield Street
Birmingham B1 3AG
England
Tel: 0121 236 9101

Arkley Sportscars
Midland House
Hayes Lane
Lye
West Midlands DY9 8RD
England
Tel: 01384 422424

Arkley USA - See North
American Arkley

Armteca Automotive
Electronics
51 Marlborough Gardens
Grange Park
Hedge End
Southampton
Hampshire SO30 2UT
England

Autocar Electrical Equipment
Ltd (Microdynamics parts)
77-88 Newington Causeway
London SE1 6BJ
England
Tel: 0171 403 5989

Automotive Components Ltd
(Compomotive Wheels)
No 4 - 6 Wulfrun Industrial
Estate
Stafford Road
WOLVERHAMPTON
West Midlands WV10 6HG
England

Auto-shop (See Hurricane
Racing Engines)

Autostorica
PO Box 355
Woking
Surrey GU22 9QE
England
Tel: 01539 734313

AVO UK/Chassis Dynamics
Unit 108
Lawrence Leyland Industrial
Estate
Erthlingborough Road
Wellingborough NN8 1RA
England
Tel: 01933 270504

AWF
Gratton House
Gratton Street
Cheltenham
Gloucestershire GL50 2AS
England

Tel: 01242 228111

Barrett Enterprises
Div 165
2060 Avenida De Los Arboles
Thousand Oaks
CA 91362-1361
USA
Tel: 805 494 6847

T. F. Bell & Co (Insurance
Brokers) Ltd
39 Station Road
Hinkley
Leicestershire LE10 1AP
England
Tel: 01455 251199

BGH Geartech
Unit 3
Little Telpits Farm
Woodcock Lane
Grafty Green
Lenham
Maidstone
Kent ME17 2AY
England
Tel: 01622 851140

Brembo Discs - see EC Parts
Ltd

British Motor Industry Heritage
Trust
Heritage Motor Centre
Banbury Road
Gaydon
Warwickshire CV35 0BJ
England
Tel: 01926 641188

British Wire Wheel
1600 Mansfield Street
Santa Cruz
USA

Brown and Gammons Ltd
18 High Street
Baldock
Herts SG7 6AS
England
Tel: 01462 490049

Carcraft
Motor Engineers & Restorers
Nacton Road Industrial Estate
Ipswich
Suffolk IP3 ORR
England
Tel: 01473 723991

Carr Reinforcements Ltd
(Carbon Fibre seats and panels)
Unit 1A
Heapriding Business Park
Ford Street
Chestergate
Stockport
Cheshire SK3 OBT
England
Tel: 0161 429 9380

Classic Automobiles of Rutland
Melton Mowbray
England
Tel: 01664 65936

Corbeau USA
9503 South
560 West Sandy
Utah 84070
USA
Tel: 801 255 3737

Crane Cams (UK)
PO Box 33
Welshpool
Powys SY21 9ZZ
Wales
Tel: 01938 556614

Corbeau Seats (Protech
Seating Ltd)
Ivyhouse Industrial Area
Hastings
East Sussex
England TR35 4NN
Tel: 01424 435480

Cosmos Trading Co. Inc
20-7 Nihonbasi Kodenmacho
Tokyo
Japan

Crane Cams N.V. (Europe)
Autolei 151

2160 Wommelgem
Belgium
Tel: 3 366 4040

Dellow Automotive
37 Daisy St
Revesby
Sydney
Australia 2212
Tel: 2 774 4419

Dunlop Australia
South Pacific Tyres Limited
Private bag 12
Cambellfield 3061
Victoria
Australia
Tel: 33050333

Dunlop (New Zealand)
South Pacific Tyres NZ Limited
PO Box 40343
Upper Hutt
New Zealand
Tel: 45288009

Dunlop Tyres (Canada) Limited
260 Hanlan Road
Woodbridge
Ontario L44 3P6
Canada
Tel: 68515784

Dunlop UK
Motorsport Division
SP Tyres UK Limited
Fort Dunlop
Birmingham B24 9QT
England
Tel: 0121 306 3516

Dunlop (USA) Tire Corporation
PO Box 1109
Buffalo
New York 14240
USA
Tel: 716 773 8218

EC Parts Ltd (Brembo)
Unit C2 Lincoln Park
Buckingham Road Industrial
Estate
Brackley

Northants NN13 5BE
England
Tel: 01280 700664

Facet Enterprises Inc.
Automotive Components
Marketing
Elmira
NY 14903
USA
Tel: 607 737 8243

Farndon Engineering
Bayton Road
Exhall
Coventry CV7 9EJ
England
Tel: 01203 366910

Faspec
1036 SE Stark Street
Portland
OR 97214
USA
Tel: 503 232 1232

Fibresports
34-36 Bowlers Croft
Cranes Industrial Area
Basildon
Essex SS14 3ED
England
Tel: 01268 527331/282723

FlowTech Racing Ltd
Unit One Rutherfords Business
Park
Marley Lane
Battle
East Sussex TN33 0RD
England
Tel: 01424 772202

Frogeye Car Company Ltd
Simeon Motor Works
Simeon Street
Ryde
Isle of Wight PO33 1JQ
Tel: 01983 616616

Frontline Spridget Ltd
Unit 1 Venture House
Melcombe Road

Oldfield Park
Bath
Avon BA2 3LR
England
Tel: 01225 446544

Fuel System Enterprises Ltd
180 Hersham Road
Hersham
Walton on Thames
Surrey KT12 5QE
England
Tel: 01932-231973

Gathercole Racing
Unit 1A Ark Farm
Wood Lane
Ramsey
Huntingdon PE17 1UA
England
Tel: 01487 815807

Gillspeed
48 Regent Street
Oakleigh
Victoria
Australia 3166
Tel: 3 9568 0688

GMA Sas (for MWS wheels)
Di Gallorni Marco
27100 Pavia
Italy

Goodridge (Benelux)BV
Leeuwenstein 40 2627 AM
Delft
The Netherlands
Tel: 15 565232

Goodridge (Espana)SL
Plaza Pedagoga Raquel Paya, 8
46006 Valencia
Spain
Tel: 6 333 2044

Goodridge (France) Sarl
22 BD Victor Hugo 77000
Melun
France
Tel: 1 64 38 44 44

Goodridge (Japan) Ltd

35-15(701) Kamikitazawa 4-
Chrome
Setagaya-ku
Tokyo
156 Japan
Tel: 3 3329 5504

Goodridge (UK) Ltd
Exeter Airport Business Park
Exeter
Devon EX5 2UP
England
Tel: 01392 69090

Goodridge (USA) Inc
1880 Del Amo Blvd, Unit D
Torrance
CA 90501
USA
Tel: 213 533 1924

Hardy Engineering - Gearboxes
268 Kingston Road
Leatherhead
Surrey
England
Tel: 01372 378927

Hobbs Sport Ltd (Goodridge
Hose)
Unit 4D
Brentmill Industrial Estate
South Brent
Devon TQ10 9YT
England
Tel: 01364 73956

Huffaker Racing
Sears Point Raceway
28013 Arnold Dr
Sonoma
CA 95476
USA
Tel: 707 935 0533

Hurricane Racing Engines
Auto-Shop Haarlem B.V.
Gonnetstraat 1- 3
2011 KA Haarlem
The Netherlands
Tel: 23 310306

Induction Technology Group

Ltd
Unit 5 Fairfield Court
Seven Stars Industrial Estate
Whitley
Coventry
West Midlands CV3 4LJ
England
Tel: 01203 305386

Janspeed Engineering Ltd
Castle Road
Salisbury
Wiltshire SP1 3SQ
England
Tel: 01722 321833

J.E.M. (obsolete brake parts)
Ashcroft House
Druid Street
Hinkley
Leicestershire LE10 1QH
England
Tel: 01455 230626

Kenlowe Ltd (electric fans)
Burchetts Green
Maidenhead
Berkshire SL6 6QU
England
Tel: 0162 882 3303

Kent Performance Cams Ltd
Units 1-4 Military Road
Shorncliffe Industrial Estate
Folkestone
Kent CT20 3SP
England
Tel: 01303 248666

Kool Louvres
14 Walton Way
Aylesbury
Buckinghamshire HP21 7JL
England
Tel: 01296 88548

Lifeline Fire and Safety Ltd
1 Portway Close
off Torrington Avenue
Coventry
West Midlands CV4 9UY
England
Tel: 01203 471207

Richard Longman & Co
5 Airfield Road
Airfield Way Industrial Estate
Christchurch
Dorset BH23 3TG
England
Tel: 0202 486569

Long Motor Corporation - see
Victoria British Ltd

Luke Racing Systems Ltd
Unit 6 Diplocks Buildings
Diplocks Way
Hailsham
Sussex
England
Tel: 01323 844791

Marshall of Cambridge
Jesus Lane
Cambridge
Cambridgeshire CB5 8BH
England
Tel: 01223 62211

Peter May Engineering. Ltd
Midland House
Hayes Lane
Lye
West Midlands DY9 8RD
England
Tel: 01384 422424

Mike Meeks (secondhand
spares)
35 Warwick Road
Boscombe
Bournemouth
Dorset BH7 6JR
England
Tel: 01202 433233
(open Wednesdays &
Saturdays)

Merlin Motorsports
Castle Coombe Circuit
Chippenham
Wiltshire SN14 7EX
England
Tel: 01249 782101

Merryman Modifications

2461 Carlisle Pike
Hanover
PA 17331
USA
Tel: 717 633 6083

Merv Plastics (cables and wiring
accessories)
201 Station Road
Beeston
Nottingham
Notts NG9 2AB
England
Tel: 0115 9222783

Metro Products Ltd (Stoplock)
Eastman House
118 Station Road East
Oxted
Surrey RH8 0QA
England
Tel: 01883 717644

M&G International
International House
Lord Street
Birkenhead
Liverpool L41 1HT
England
Tel: 0151 666 1221/1666

MG Parts Centre
c/o Bill Richards
5 Alden Terrace
Howell
New Jersey 07731
USA

Mini Classic Cars (E. Hankins)
165A Southwood Road
Hayling Island
Hants PO11 9PY
England

Mini Mania West
31 Winsor Street
Milpitas
CA 95035
USA
Tel: 408 942 5595

Mini Mania East
1895-F Beaver Ruin Rd

Norcross
GA 30071
USA
Tel: 404 263 6595

Mini Spares (Northern) (Mini
Mania parts from the USA)
22 Freeman Way
Harrogate Business Park
Wetherby Road
Harrogate
North Yorkshre HG3 1PH
England
Tel: 01423 881800

Mini Spares Centre Ltd
29-31 Friern Barnet Road
London N11 1NE
England
Tel: 0181 368 6292

Minilite
Two Eversley Way
Thorpe Industrial Estate
Egham
Surrey TW20 4RG
England
Tel: 01784 433446

Minor Mania Ltd
1/3 Hale Lane
Mill Hill Broadway
Mill Hill
London NW7 3NU
England
Tel: 0181 959 0818

MK Parts
Braun Mattstr 2
D76532 Baden Baden
Germany
Tel: 722 11300

Morris Minor Centre
(Birmingham) - Sierra gearbox
conversion
Birmingham Parade
2 Camden Street
Birmingham B1 3BN
Tel: 0121 236 1341
Moss Motors Ltd
7200 Hollister Avenue

PO Box 847
Goleta
CA 93116
USA
Tel: 805 968 1041

Moss Sprite & Midget B, C, V8
Centre,
22-28 Manor Road
Richmond
Surrey TW9 1YB,
England
Tel: 0181 948 6665

Mota Lita (USA)
503 Corporate Sq
1500 NW 62nd St
Fort Lauderdale
FL 33309
USA
Tel: 305 776-BRIT

Motobuild Ltd
328 Bath Road
Hounslow
Middlesex TW4 7HW
England
Tel: 0181 570 5342/572 5437

Motor Wheel Service (MWS)
Southern Region
Langley Business Park
Station Road
Langley
Berkshire SL3 6EP
England
Tel: 01753 549360

Motor Wheel Service (MWS)
Northern Region
Shentonfield Road
Sharston
Wythenshawe
Manchester M22 4RW
England
Tel: 0161 428 7773

Mr Fast'ner
Dept C6
Units 1 & 2
Warwick House Industrial Park
Banbury Road
Southam

Warwickshire CV33 0HL
England

MWS/Leberfinger
Kiebitzhorn 31
2000 Hamburg-Barsbuttel
Germany

North American Arkley
PO Box 18667
Asheville
NC 28814
USA
Tel: 704 628 9626

Northwest Import Parts
10915 S.W. 64th Avenue
Portland
OR 97219
USA
Tel: 503 245 3806

Nottingham MG Centre
Unit 13-14
Colwick Business Park
Private Road No 2
Colwick
Nottingham NG4 2JR
England
Tel: 0115 9615283

Octagon Spares BV
Mr G de Groot
Burg Van Niekerklaan 19
2182 GK Hillegom
Holland

Octagon Workshop Ltd
Main Road
Union Mills
Isle of Man
Tel: 01624 852220

O/D Spares Overdrive
specialists
Unit A2 Wolston Business Park
Main Street
Wolston
Nr Coventry
West Midlands CV8 3FU
England
Tel: 01203 543686

Omni-Auto's Nik Handford
23 Cranham
Yate
Bristol BS17 4JT
England
Tel: 01454 313600

Perfect Nostalgia
The Workshop
Rook Farm
Rotherwick
Hants RG27 9BJ
England
Tel: 01256 768678

Pro-Align
30 Ross Road Business Centre
Northampton
Northamptonshire NN5 5AX
England
Tel: 01604 588880

Project Company
7401 Eastmoreland Road
Suite 515
Annandale
VA 22003
USA
Tel: 703 658 0253

Quaife Power Transmissions
R.T. Quaife Eng Ltd
Vestry Road
Otford
Sevenoaks
Kent TN11 0LQ
England
Tel: 01732 741144

Race Techniques - Engine
Development
Unit 9/10 Elim Works
Dunalley Parade
Cheltenham
Gloucestershire GL50 4LS
England
Tel: 01242 245640

Reco-Prop
Unit 4
Newtown Trading Estate
Chase Street
Luton

Beds LU1 3QZ
England
Tel: 01582 412110

Ridgard Seats
Plandevice Ltd
Parsons Farm
Tilbury Road
Ridgewell
Essex CO9 4RL
England
Tel: 01440 788141

Ripspeed International
54 Upper Fore Street
Edmonton
London N18 2SS
England
Tel: 0181 803 4355

Ron Hopkinson MG Parts
Centre
850 London Road
Derby
Derbyshire DE2 8WA
England
Tel:01332 756056

Safety Devices
Regal Drive
Soham
Cambridgeshire CB7 5BE
England
Tel: 01353 624624

Schofield & Samson Ltd
Tuscany Wharf
4b Orsman Road
London N1
England
Tel: 0171 739 6817/8

Serck Marston (Competition
Products Division)
5-7 Cullen Way
Park Royal Road
London NW10 6LY
England
Tel: 0181 965 5442

Serck Marston
Heming Road
Washford Industrial Estate

Redditch
Worcestershire B98 0DZ
England
Tel: 01527 510111

Seven Enterprises Ltd
716 Bluecrab Road
Newport News
VA 23606
USA
Tel: 804 873 0007

SPA Design Ltd
The Boat House
Lichfield Street
Fazely
Tamworth
Staffordshire B78 3QN
England
Tel: 01827 260528

Spax Ltd
Telford Road
Bicester
Oxfordshire OX6 OUU
England
Tel: 01869 244771

Speedograph Richfield Ltd
Rolleston Drive
Arnold
Nottingham,
Notts NG5 7JR
England
Tel: 0115 9264235

Speedwell Engineering
1711 First St
San Fernando
CA 91340
USA
Tel: 818 898 9151

Sports & Classics
512 Boston Post Road
Darien
Conn 06820
USA
Tel: 203 655 8731

Stack Inc
501 Wesley Plantation Dr. NW
Duluth
GA 30136
USA

Tel: 404 844-8801

Stack Limited (Tachometer/
revcounter)
Wedgewood Road
Bicester
Oxfordshire OX6 7UL
England
Tel: 01869 240404

Stockbridge Racing (Willans
Safety Harness)
Grosvenor Garage
Stockbridge
Hampshire
England
Tel: 01264 810712

Toyota Sport
Toyota (GB) Ltd
Technical Centre
8 Steer Place
Bonehurst Road
Salfords
Redhill
Surrey RH1 5EF
England

Tel: 01293 774866

Victoria British Ltd
PO Box 14991
Lenexa
Kansas 66285-4991
USA
Tel: 913 541 1525

Willans - see Stockbridge
Racing

Winners Circle
19144 Detroit Road
Rocky River
Ohio 44116-1722
USA
Tel: 216 333 4666

Wolseley 1500 Spares
2 Purberry Grove
Ewell
Surrey KT17 1LU
England
Tel: 0181 393 2194/0860
360690.

Appendix II
Race Series (British & European)

There are several race series open to the Midget/Sprite as well as classes in hillclimbing, sprinting and rallying.

Each type of event will have general and safety regulations for cars and specific regulations for classes (classes are divided into categories of near standard, modified classes and in-betweens). Separate race series for the Midget/Sprite are run by the Austin Healey Club, MG Car Club and MG Owners Club and rallying comes under the umbrella of the Historic Rally Register. Hillclimbing and sprinting events are covered by the Hillclimb and Sprint Association.

The Austin Healey Club has two main classes for circuit racing - standard and modified. Both classes are fairly free but becoming less so and, like any class, are subject to yearly change. Standard class cars are more relaxed than the MGCC standard class. Modified cars are very free, although, in recent years, have reverted from slicks to a control tyre which may be more to do with letting the big Healeys catch up than an attempt to limit the expenses of modified racing.

The MG Car Club has three classes:

A, B and C, which are fully modified, race modified and road-going cars respectively. General modifications for all classes allow removal of bumpers and interior trim and for a racing seat to be fitted. The extent of bodywork modifications varies from class to class with limits for all classes on air dams and spoilers and the number of glass fibre panels that can be used. Hardtops are permitted.

For any series the most detailed regulations concern the engine and the MGCC series is no exception. It's easier to detail some of what is not allowed rather than what is: the original block and head must be used in all classes, thus outlawing 8 port or 16 valve heads and engine transplants. Forced induction is also illegal, as are crank/flywheel triggered ignition/management systems. Carburation is restricted in class C to original equipment carbs and manifold size.

Suspension modifications vary from class to class, with no telescopic absorber conversions front or rear, or both, dependent on class. Class C does not permit anti-tramp bars and panhard rods.

Limited slip diffs are not permitted in

classes B and C and close ratio gears not allowed in class C. For all classes the gearbox must be the original type with only four forward gears.

Braking modifications in class C exclude larger discs, drums, aluminium drums and bias valves. Class B is only restricted to a front disc/rear drum layout. Wheel rim width is limited to 4.5ins in class C and 6ins in class B. Diameter is limited to 13ins in any class. Alloy wheels are not permitted in class C. Tyres are free for class A and restricted to listed road tyres for class B. Class C tyres are limited to a single make - Bridgestone.

There are minimum weights of 1279lbs (580kg) and 1500lbs (680kg) for classes B and C with 1526lbs (692kg) minimum for 1500cc-engined cars.

For a more detailed look at the regulations the club will provide a full set on request, but the summary above outlines the main points which give a feel for the differences between classes to help you decide which class to compete in. For instance, any half-decent modified road car is most likely to require de-modifiying for class C.

The MG Owners Club series is geared to road-going cars and all cars must be road legal. Minimum weights are 1478lbs (670kg) for A-series-engined cars and 1555lbs (705kg) for 1500 Midgets. Bodywork modifications are very limited and, for some reason, louvred bonnets are not permitted. One-piece GRP fronts are not allowed so I guess carbon fibre would be okay (just kidding!). Engine regulations are fairly tight, as are allowed suspension modifications. The overall limitation on the series must be the tyre size of 155 x 70. Generally, this series looks to be the most restrictive and probably therefore the cheapest. The MGOC provides a useful starter's information pack for would-be entrants as well

European Series

Slightly different to other series is the Dutch or, more accurately, European series for Sprites. Organised by Dutchman Peter Bekker, it's a series with a professional and businesslike structure. Instead of paying to compete, the series is free if the driver races in six out of the seven races. Drivers get a free set of control tyres (Goodyear GT2 155/70) and also free transportation via Ferry Masters. Once at the race track the whole family is catered for (a social programme is considered important to the series, which is televised on Eurosport). Racing is at a variety of venues and the series is planned to continue until at least the year 2000. The regulations are based on FIA rules but allow the use of 1275cc engines instead of 1000cc engines. On the whole, I would say that it might well be cheaper, even with European travel, to compete in this series rather than in the UK.

SPEEDPRO SERIES

AMERICAN/ENGLISH GLOSSARY OF AUTOMOTIVE TERMS

American	English
A-arm	Wishbone (suspension)
Antenna	Aerial
Axleshaft	Halfshaft
Back-up	Reverse
Barrel	Choke/venturi
Block	Chock/wedge
Box end wrench	Ring spanner
Bushing	Bush
Clutch hub	Synchro hub
Coast	Freewheel
Convertible	Drop head
Cotter pin	Split pin
Counterclockwise	Anti-clockwise
Countershaft	Layshaft (of gearbox)
Crescent wrench	Open-ended spanner
Curve	Corner
Dashboard	Facia
Denatured alcohol	Methylated spirit
Dome lamp	Interior light
Driveaxle	Driveshaft
Driveshaft	Propeller shaft
Fender	Wing/mudguard
Firewall	Bulkhead
Flashlight	Torch
Float bowl	Float chamber
Freeway, turnpike, etc.	Motorway
Frozen	Seized
Gas tank	Petrol tank
Gas pedal	Accelerator pedal
Gasoline (gas)	Petrol
Gearshift	Gearchange
Generator (DC)	Dynamo
Ground	Earth (electrical)
Header/manifold	Manifold (exhaust)
Heat riser	Hot spot
High	Top gear
Hood	Bonnet (engine cover)
Idle	Tickover
Intake	Inlet
Jackstands /Safety stands	Axle stands
Jumper cable	Jump lead
Keeper	Collet
Kerosene	Paraffin
Knock pin	Roll pin
Lash	Freeplay/Clearance
Latch	Catch
Latches	Locks
License plate /tag plate	Number plate
Light	Lamp
Lock (for valve spring retainer)	Split cotter (for valve cap)
Lopes	Hunts
Lug nut	Wheel nut
Metal chips or debris	Swarf
Misses	Misfires
Muffler	Silencer
Oil pan	Sump
Open flame	Naked flame
Panel wagon/van	Van
Parking light	Sidelight
Parking brake	Handbrake
Piston pin or wrist pin bearing/bush	Small (little) end bearing
Piston pin or wrist pin	Gudgeon pin
Pitman arm	Drop arm
Power brake booster	Servo unit
Primary shoe	Leading shoe (of brake)
Prussian blue	Engineer's blue
Pry	Prise (force apart)
Prybar	Crowbar
Prying	Levering
Quarter window	Quarterlight
Recap	Retread
Release cylinder	Slave cylinder
Repair shop	Garage
Replacement	Renewal
Ring gear (of differential)	Crownwheel
Rocker panel	Sill panel
Rod bearing	Big-end bearing
Rotor/disk	Disc (brake)
Secondary shoe	Trailing shoe (of brake)
Sedan	Saloon
Setscrew, Allen screw	Grub screw
Shift fork	Selector fork
Shift lever	Gearlever/gearstick
Shift rod	Selector rod
Shock absorber, shock	Damper/shocker
Snap-ring	Circlip
Soft top	Hood
Spacer	Distance piece
Spare tire	Spare wheel
Spark plug wires	HT leads
Spindle arm	Steering arm
Stablizer or sway bar	Anti-roll bar
Station wagon	Estate car
Stumbles	Hesitates
Tang or lock	Tab
Taper pin	Cotter pin
Teardown	Strip(down)/dismantle
Throw-out bearing	Thrust bearing
Tie-rod (or connecting rod)	Trackrod (of steering)
Transmission	Gearbox
Troubleshooting	Fault finding/diagnosis
Trunk	Boot
Tube wrench	Box spanner
Turn signal	Indicator
Valve lifter	Tappet
Valve lifter or tappet	Cam follower or tappet
Valve cover	Rocker cover
VOM (volt ohmmeter)	Multimeter
Wheel cover	Roadwheel trim
Wheel well	Wheelarch
Whole drive line	Transmission
Windshield	Windscreen
Wrench	Spanner

Index

SPEEDPRO SERIES